P9-CQG-098

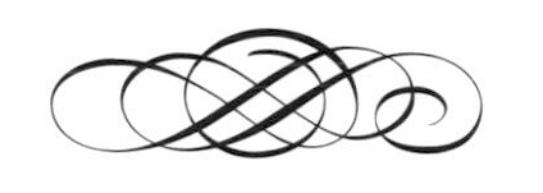

WYLLIE, ACTON
& GOLDBLATT'S

THE TIME TRAVELER'S HANDBOOK

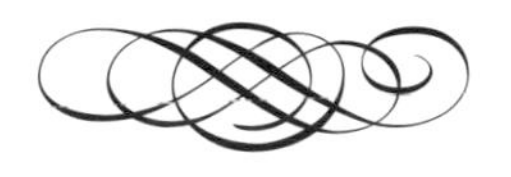

18 EXPERIENCES FROM THE
ERUPTION OF VESUVIUS TO WOODSTOCK

THE TIME TRAVELER'S HANDBOOK

18 EXPERIENCES FROM THE
ERUPTION OF VESUVIUS TO WOODSTOCK

RESEARCHED AND WRITTEN BY

JAMES WYLLIE, JOHNNY ACTON
& DAVID GOLDBLATT

First published in Great Britain in 2015 by
Profile Books, London WC1X 9HD
www.profilebooks.com

THE TIME TRAVELER'S HANDBOOK

Published in 2016 by
Harper Design
An Imprint of HarperCollins*Publishers*
195 Broadway
New York, NY 10007
Tel: (212) 207-7000
Fax: (855) 746-6023
harperdesign@harpercollins.com
www.hc.com

Distributed throughout the world by
HarperCollins*Publishers*
195 Broadway
New York, NY 10007

ISBN 978-0-06-246939-7
Library of Congress Control Number 2015955962

Printed in Italy

First Printing, 2016

Typeset in Caslon, Sancreek, and Copperplate Gothic
to a design by Henry Iles.

Thanks to Sally Holloway, Sonia Land, Karim Noorani,
Nikky Twyman, Henry Iles, Dominic Beddow, and Mark Ellingham.

Contents

EPIC JOURNEYS & VOYAGES

EXTREME EVENTS *

* These events are not covered under the company's insurance policy.

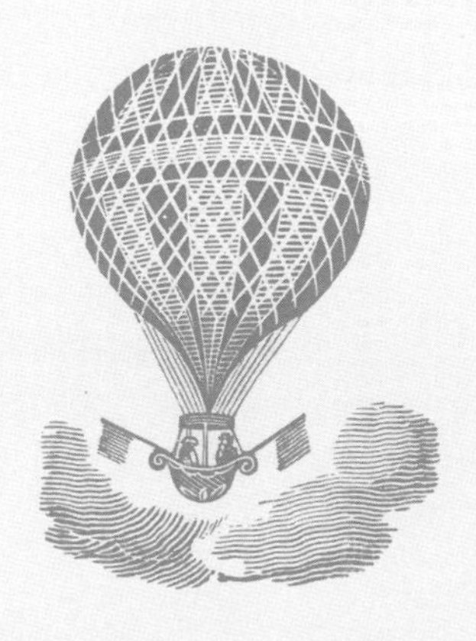

WYLLIE, ACTON
& GOLDBLATT'S

TIME TRAVEL TOURS

"History repeats itself. First time as tragedy, second time as vacation."

SOME SAY THE PAST IS A FOREIGN COUNTRY. We say get your passport. At WYLLIE, ACTON & GOLDBLATT (WAG) TIME TRAVEL TOURS, we take you back, set you down, and bring you home from some of the finest moments in human history. With our unique Chronoswoosh™ time exchange plasma shuttle technology, we not only offer the most accurate return to the past, but minimal interference with the time-space continuum. No more getting abandoned in the wrong century, no more returning to find you are your great aunt.

At WAG we believe that history the second time around need not be farce but a celebration, a party, and a vacation. If that sounds like your idea of a good time past, then our CELEBRATIONS & EXHIBITIONS suite of trips is for you. Try the late feudal pomp of the FIELD OF THE CLOTH OF GOLD, where the nobilities of France and England gathered in great alfresco camps to drink, feast, joust, and doff their caps to King Henry VIII and King Francis I. The GREAT EXHIBITION OF 1851 certainly deserves another look and we give you the opportunity to explore Victorian London and the cornucopian contents of the Crystal Palace. For the more hedonistic, VE DAY IN LONDON 1945 and the 1969 WOODSTOCK FESTIVAL in upstate New York offer contrasting experiences of collective ecstasy.

For more experienced time travelers we recommend moving on to an event that has really made a difference. Our carefully curated selection of MOMENTS THAT MADE HISTORY show the wheels of change in motion while delivering the most extraordinary sensory experiences. Feel the ancien régime crumble and the modern world emerge as CHARLES I IS EXECUTED at the end of the English Civil War or MARCH WITH THE WOMEN OF PARIS in the heat of the French Revolution. For those with more contemporary tastes we offer the poles of the short twentieth century: the ASSASSINATION OF ARCHDUKE FERDINAND IN SARAJEVO that triggered the First World War in 1914; and the FALL OF THE BERLIN WALL in 1989 that brought the long European struggle to a close.

For a more reflective journey to the past, WAG also offers a great treasure trove of CULTURAL & SPORTING SPECTACULARS. Our selection of classic moments takes you to events that seemed unrepeatable and offers them up in all their replayed glory. From the spectacle of the ANCIENT OLYMPICS to the opening night of SHAKESPEARE'S GLOBE, from the BIRTH OF BEBOP in war-time New York City to the invention of the BEATLES IN HAMBURG, or Muhammad Ali fighting George Foreman in THE RUMBLE IN THE JUNGLE, there is something to suit all tastes.

For the more adventurous and stoical of our travelers, we have recently developed the long-form EPIC JOURNEYS & VOYAGES: six months with MARCO POLO IN XANADU in thirteenth-century China; or the chance to crew on the THREE-YEAR VOYAGE OF CAPTAIN COOK to Australia. For those with the strongest nerves and most balanced dispositions we are pleased to include our EXTREME EVENTS: the ever-popular ERUPTION OF VESUVIUS and cataclysmic destruction of Pompeii, the late-medieval madness of the PEASANTS' REVOLT that set London alight, and our newest offer, a front-seat view of our first American Civil War trip, the hubris and the chaos of the FIRST BATTLE OF BULL RUN.

Whatever trip you choose, the TIME TRAVELER'S HANDBOOK is your Owl of Minerva. In putting together this guide we've been where you are going, we've flown home at dusk, and, at last, with the wisdom of hindsight actually available to us, charted a course through the past on your behalf. We really will take you to the right place at the right time, every time. Alongside the key moments, we let you know WHERE TO STAY, WHAT TO EAT, and HOW TO PAY, and of course HOW TO GET HOME. Our Uncle Karl thought that "Man made history but not in circumstances of his own choosing." We can't let you make history—it's already a done deal—but we can let you choose the circumstances. Welcome to the past; we'll take you there.

THE SMALL PRINT: TERMS AND CONDITIONS OF TRAVEL

APPEARANCE AND DRESS CODE

☞ It is important to blend in with the crowd as far as possible, and for this reason WAG will provide you with suitable attire for all of your journeys. You may, however, already possess appropriate clothing or want to try making your own. If so, please read the guidance notes carefully. In cases where having a skin tone and physiognomy markedly differ from that of the locals would attract undue attention, WAG's experienced prosthetic and make-up department will swing into action.

HEALTH ISSUES

☞ All travelers will be subject to stringent HEALTH CHECKS ahead of their journey. An innocuous-seeming virus could have devastating consequences for a historical community which has not developed immunity to it. For similar but reverse reasons, travelers to high-risk destinations will be subject to health checks on their returns, and quarantine will be enforced if necessary. Some trips are more strenuous than others, the long voyages and extreme events in particular. We reserve the right to refuse time travel to clients whose medical condition renders them vulnerable. We advise all travelers to take out appropriate backdated medical insurance.

LANGUAGES AND COMMUNICATION

☞ Travelers should assume that in all but the most contemporary journeys their mother tongue will be of limited use. English speakers who go unprepared to Elizabethan London will find it hard to understand others or to make themselves understood. Consequently WAG provides a basic grounding in the LOCAL LANGUAGE of your journey and an introductory course to body language, ritual, customs, and deportment. The quoted price of each trip includes a two-day residential preparation and orientation programme. Please note: these courses are compulsory and a minimal standard of competence is required of all travelers.

THE TIME-SPACE CONTINUUM

All of our trips have been carefully selected so that you can blend into the past without drawing special attention to yourself. (It is for this reason, of course, that we cannot offer moon landings or intimate liaisons with the likes of Caesar or Napoleon.) Even so, you must exercise great care not to do anything that might interfere with the events you have chosen to visit. The time–space continuum is sufficiently robust to cope with your presence and your minor interactions with the events of the past, but grand gestures are simply not on. WAG reserves the right to transport any customer who appears poised to infringe this regulation instantly back to the present, with no compensation payable.

Minor interactions with other members of the crowd are permitted, as are everyday activities such as EATING AND DRINKING. Any changes such interactions make to the course of history will be negligible and tolerable (e.g., you may return to find that you like speed metal rather than jazz, or that your partner is called Lionel rather than Pam). Such alterations must be regarded as an occupational hazard. It goes without saying that bringing back SOUVENIRS is not permitted, as this would play havoc with the market for antiques. Strictly no mobile phones and no cameras. Don't even think about selfies.

WAG will accept no responsibility for the consequences of failing to adhere to these terms and conditions.

PART ONE

Celebrations & Exhibitions

The Field of the Cloth of Gold

JUNE 8–24, 1520 ✵ NEAR CALAIS, ENGLAND

FOR JUST OVER TWO WEEKS IN JUNE 1520, Henry VIII and Francis I, the kings of England and France, and most of their feudal nobility, gathered for a great outdoor meeting in northern France. They came together ostensibly to make peace and celebrate the betrothal of Francis's son and Henry's daughter but it was an occasion soaked in sixteenth-century realpolitik. THE FIELD OF THE CLOTH OF GOLD (as it was dubbed by eighteenth-century historians) provided an extraordinary opportunity for two great Renaissance princes to display themselves to each other and their followers as warrior kings, chivalric gentlemen, and luminous stars in Europe's political firmament.

Settled into vast and ostentatious tent cities, equipped with splendid temporary palaces, the two sovereigns first met and then joined together to stage a great knightly tournament of jousting and foot combat, punctuated by intense merrymaking, feasting, and dancing.

PLEASE NOTE that this trip is based on attendance in the English camp. For those that prefer a more Francophone experience, we hope to be able to offer *le champ de la toile d'or* experience in the near future.

BRIEFING: GRAND SUMMITS

Grand summit meetings between the French and English kings had a considerable track record. In 1254 Henry III of England met Louis IX of France in Chartres and they rode together to Paris for a great banquet. Things went so well they did it all again in 1259 and signed a peace treaty. Things were taken up a notch in scale and splendour by the 1396 meeting of Richard II and Charles VI. Held in the midst of the Hundred Years War it secured a temporary peace by marrying Richard to seven-year-old Princess Isabelle of France.

Like nearly every European monarch since the fall of the Roman Empire, Henry VIII of England and Francis I of France spent a great deal of time and money on warfare; that, after all, was the very purpose of the sovereign, to defend and expand the realm (and in his own body and demeanour express the warrior manliness of the feudal nobility). However, in the early sixteenth century there were countervailing trends. Intellectually, the new humanism of writers like Erasmus argued that monarchs' power was best expressed by keeping the peace, and the warrior creed shifted towards the restraints of the chivalric code. Politically, two of Europe's most important statesmen had reasons for wanting to halt the internecine warfare. Pope Leo X sought unity because he feared the rising power of the Ottoman Empire in the east. England's Cardinal Wolsey, who was pushing England from the periphery to the centre of European politics, sought stability.

The 1518 Treaty of London was the outcome of a short period of competitive peacemaking, with Wolsey and the Pope inviting all European states to agree to an enduring peace. The French signed up, and to seal the deal the Dauphin Francis was betrothed to Henry's daughter, Princess Mary. The small print included an agreement that the monarchs would meet and hold a tournament. The combination was essential, for the jousts and tourneys would allow both sovereigns to demonstrate warrior credentials but simultaneously show they could forsake war for chivalric reasons.

There then followed eighteen months of competitive gift-giving between the kings and pettyfogging, super-politicised debate over the details and protocols of the event, led by Gaspard de Coligny, Marshal of France, and Charles Somerset, Earl of Worcester and Lord Chamberlain. However, by April 1520 the deal was done and the senior officers of both Royal Households bent to the enormous logistical task of staging the event, transporting over 6,000 people, along with their horses and baggage, from England, and an equal number of the French court from around the country.

THE TRIP

Your POINT OF ARRIVAL on the morning of June 8, 1520 will be the very rutted road that runs between the small towns of ARDRES and GUINES (in what is now northern France). When you arrive on the path about a mile short of Guines, you will actually be in England, or rather the PALE OF CALAIS, an area of land extending about ten miles from the English port ceded by the French in 1347. You will need to return here on June 24th for your departure.

The main body of the English court, nearly 6,000 strong, will have arrived a day beforehand but the road is still likely to be busy. As you head east towards the English camp in GUINES you might see the French nobles Lord Chancellor Antoine Duprat and Admiral Bonnivet riding brusquely past you. They are paying a final courtesy call on Henry. A half dozen English nobles will, on their part, ride out to Ardres to see King Francis. There will also be a steady flow of carts and mules carrying the enormous quantity of supplies required to build and feed the English camp. Expect a selection of vagabonds, beggars, and thieves to be lurking around the fringes of the camp sensing that plentiful scraps and charity will be made available over the coming days.

THE ENGLISH CAMP

When GUINES first comes into view it will not be the town's stone castle, church spire, or cluster of housing that catches your eye, but the vast tent city of over 300 pavilions in front of it. Almost the entirety of the Tudor nobility and their retinues are here under canvas. Round, square, and rectangular tents draped in sparkling colored cloth are clustered to create instant noble houses, some with as many as a dozen tents linked by covered corridors and galleries. Glastonbury Festival it isn't—though the flags, banners, and pennants flying from every mast and central tent pole are vaguely

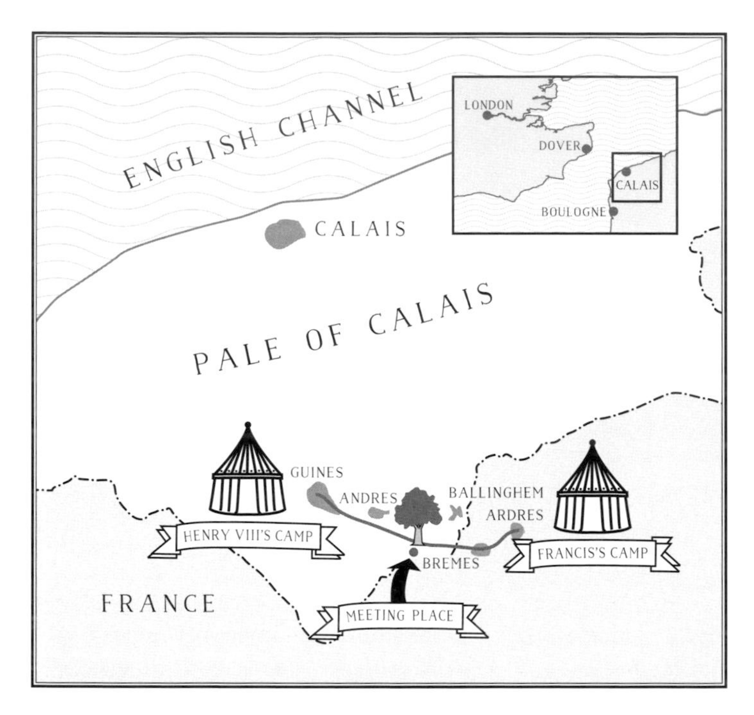

reminiscent, and around the edges is less salubrious accommodation where servants, scullery maids, armourers, and stable hands are based. Right at the heart of the site, directly before Guines Castle, is the king's Palace: a magnificent set of canvas-roofed, double-storeyed, brick buildings.

ACCOMMODATION

As you make your way down to the English camp you will be able to see that the pavilions are all trimmed with fixed and painted wooden boards, the grandest with skirting and flooring too. Tudor roses and Beaufort portcullis arms are much in evidence. Do look up, though; on top of the masts that hold up the tents there is a menagerie of beautifully carved heraldic beasts, including dragons, griffins and greyhounds, lions, stags and antelopes.

Your host for the next fortnight or so is SIR ADRIAN FORTESCUE, and you will be looking for his pavilion, marked with the Fortescue coat of arms: a blue shield with a white diagonal stripe and wavy edges, paralleled by two gold diagonal lines, or, in the Latin heraldic vernacular, *azure a bend engrailed argent cotised or*.

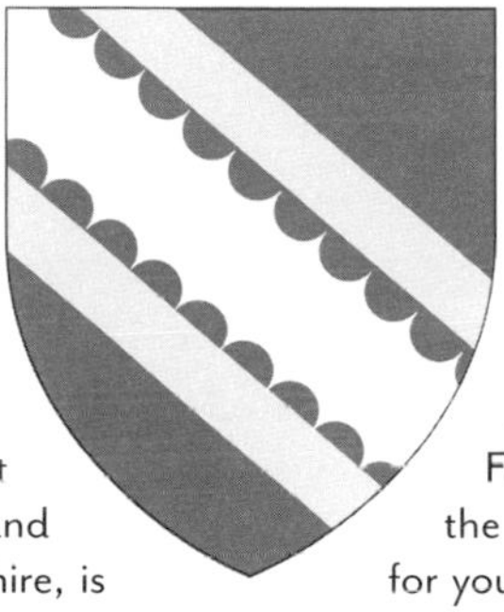

Sir Adrian, a knight of the bath with land holdings in Hertfordshire, is recently widowed and will be in deep mourning for much of the next five years. Thus distracted, he will, as long as he is not directly approached, barely notice your presence in the pavilion. Please stick rigidly to using the SMALL CIRCULAR TENT at the far end of the Fortescue pavilion. Note the Fortescue livery, here for your use, will get you into the banqueting halls, but do not expect a seat. The slop buckets are strictly your responsibility.

THE KING'S PALACE

HENRY'S PALACE at Guines, a hundred yards square, consists of four blocks around an atrium courtyard. The buildings have stone foundations, brick walls to eight feet high, then wattle and daub with clerestory windows, and an Italianate wooden cornice decorated with cross and leaf work. Above this, large timbers form a frame over which vast decorated canvas sheets are hung. The building has been completed in under three months by thousands of skilled artisans using timber and prepared wall sections from England, glass from Flanders, and cloth from all over Western Europe.

As you walk up to the ornate mock guardhouse at the front of the palace, note the splendid FOUNTAIN with its depiction of Bacchus carved from wood. On feast and banquet days this will be running with two streams of wine—one red, one white—pumped under pressure. Earthenware vessels will be available to help yourself; you won't be alone. The gatehouse itself is adorned with a statue of SAINT MICHAEL, a deliberate compliment to France's patron saint of chivalry.

Once inside the palace in the main quadrangle, you will see two sets of very ornate pavilions on either side; to your left are CARDINAL WOLSEY'S QUARTERS; to the right are those of the king's sister MARY, Duchess of Suffolk. Behind these pavilions are two long brick-walled buildings: on the left-hand side HENRY'S APARTMENTS and on the right those of QUEEN CATHERINE. The two are connected by an UNDERGROUND PASSAGE, while Henry's apartments have their very own box-tree-planted corridor leading back into Guines Castle. These tents will be closely guarded and difficult to enter, but should you get a glimpse inside, look out for Catherine's set of nine floral tapestries stitched in gold and silk. In Wolsey's main waiting chamber you can observe sumptuous TAPESTRY work, rumoured to include a set depicting the TRIUMPH OF PETRARCH. Even from outside you will smell the sweet aroma of freshly cut rushes and flowers that fill the royal Apartments.

Immediately opposite the entrance, and the building you are most likely to gain access to, is the LARGE HALL. A two-storey building, the ground floor will be a hive of activity at almost any time, housing the offices of the Lord Chamberlain, Lord Steward, and Treasure Comptroller—the three most senior positions in the royal Household. It also provides space for the royal service departments, including the warehouse and jewel house, the pantry, spicery, buttery, ewery (responsible for water and the vessels one drinks from and washes with), poultry, pitcher, and larder. The upper floor is reached by an internal staircase over which hovers a sculpture of an armoured footsoldier. Inside the BANQUETING HALL look up at the canvas roof, whose intricate decoration includes the popular if enigmatic homilies of ALEXANDER BARCLAY. He is currently a theological favourite at court, and the Benedictine monk's satirical poem, *The Ship of Fools*, published in 1513, is a good ice-breaker in conversation in 1520.

A short gallery leads out of the back of the large hall and into the PALACE CHAPEL. This large open space is decorated with gold and silver and overlooked by the two royal oratories, on the first floor; known as HOLYDAY CLOSETS, they are the box seats for the royal houses. The altar will be set with golden candlesticks and a fabulous pearl-encrusted crucifix over four feet high. Note that the English have brought their own church organ with them. Mass is held daily.

EATING AND DRINKING

Catering at the Field of the Cloth of Gold is a very stratified affair. While you may be able to gain entrance to some of the major BANQUETS to be held over the fortnight, you will for the most part be eating in the FORTESCUE PAVILION. For proper nobles and their households most meals will be self-catered. Expect a lot of pottage, broth, and tough black bread. Keep your ears open, as there will be a certain number of meals communally provided for gentlemen and perhaps their staff. Camp followers, beggars, lepers, and locals trying their luck will be ever-present around the camp's kitchens.

Water is strictly for washing in. The main drink available in the English camp is ALE, thrice-brewed in ever declining quality. The third brew, called SMALL BEER, is the least alcoholic of the ales, and consumed by servants and children. WINE, as noted previously, will be freely available on feast days but harder to come by at other times.

You will observe at the banquets that courses consist of a multitude of DISHES, both savoury and sweet. There will be a great deal of poultry, game birds, baked and steamed fish, pies and potted animals, and above all roast meats. Exotic birds, such as swans, may well be served and even reconstituted after cooking with head, neck, and feather to create an edible sculpture. The English court's account will later record that 6,475 birds of various kinds were purchased and nearly 100,000 eggs, not to mention 3,406 sheep and lambs, 842 veal calves, and 373 oxen.

THE FRENCH CAMP

It would be a considerable breach of protocol for lowly members of Sir Adrian Fortescue's household to enter the FRENCH CAMP and we advise you strongly not to. However, from a safe distance it is perfectly fine to take in the view. The French court will have made camp in the small town of ARDRES, five miles east and slightly south of the English camp and reached by the same road on which you arrived. You will see that the king's orders to build a vast ditch around the encampment and to repair the town's walls have been accomplished and the two together mark out a camp site filled with nearly 400 pavilions and a considerable swathe of the French peerage and nobility.

FRANCIS I BY JEAN CLOUET—AT AROUND THE TIME OF THE KING'S MEET. THE MEDALLION IS SAINT MICHEL. THE DRESS IS CLASSIC RENAISSANCE.

The KING'S APARTMENT should just be visible on the west side of the town, where a whole series of town houses have been combined with large pavilions to create his temporary palace. Of particular note is the huge BANQUETING HOUSE at the foot of the town, connected by a long gallery to the palace. The English knight Edmund Hall will describe it as "house of solas and sporte, of large and mightie compass." Built like a multistorey amphitheater, with three levels of stone wall, the interior is lined with balconies and, as Hall will report, decorated with "frettes and knottes made of Ive buishes and other thynges that longest would be grene for pleasure." The canvas roof is set with golden stars.

The transformation of Ardres has been overseen by the Marshal of France, GASPARD DE COLIGNY, and Grand Master of the royal Artillery, JACQUES DIT GALIOT DE GENOUILLAC, whose logistical triumphs in the past include shipping the French cannon across the Alps and winning a great victory at Milan in 1513. Most of the work on the camp has been done in Tours, over 300 miles away, where the local textile industries have provided enough skilled labour to service the enormous order from the French court. For over three months thousands of cloth, silk, and leather workers have been stitching and assembling the camp's pavilions, working in shifts around the clock in the palace of the Archbishop of Tours. Among the most eye-catching of these is the main tent of the royal Household, draped in golden cloth with three wide stripes of blue velvet, themselves encrusted with golden fleurs-de-lis. Atop this, balanced on a golden ball you should be able to make out a six-foot-high gold-and-blue wooden statue of Saint Michael holding a lance and shield, a slain serpent at his feet. Many of the other tents in the encampment feature a large golden apple carved from walnut instead. Queen Claude's pavilion close by can be picked out for its lighter, brighter golden cloth. The queen Mother's tents feature violet and crimson material.

DAY BY DAY

THURSDAY, JUNE 8TH: THE KING'S MEET

You should arrive mid-morning on Thursday, when preparations in the camp will be at fever pitch as horses are brushed and tack polished amid an air of nervous, growing anticipation. At

around 5pm you will hear three cannons, signalling the departure of Henry's entourage to THE KING'S MEET. Soon after this the distant sound of three cannons being fired in Ardres announces that Francis and his party are on their way.

Henry's entourage will be led out by one hundred archers drawn from the royal Guard and Wolsey's personal guard. They will be followed on horse by a large retinue of household gentlemen and knights. SIR ADRIAN FORTESCUE, if he manages to make it, will be in with this pack. Then you will see the senior nobility, and in their midst the king. HENRY will be preceded by the MARQUESS OF DORSET holding upright the sword of state; to his left CARDI-

HENRY VIII ARRIVES AT THE ENGLISH CAMP. NOTE THE ROYAL PALACE AND WINE FOUNTAIN ON THE RIGHT, TOURNAMENT FIELD (TOP RIGHT), AND THE DRAGON, WHICH WILL APPEAR AT MASS ON THE FINAL DAY.

NAL WOLSEY in brilliant crimson-red silk; to his right SIR HENRY GUILDFORD, Master of Horse, leading a spare mount for the king. The king's back is covered by a dozen young henchmen and hunting mates and then another tranche of senior nobles escorting bishops and foreign ambassadors. Finally come musicians and stewards: twelve mace-bearers, twelve trumpeters, then twelve heralds, all in Tudor livery. Expect plenty of MUSIC en route.

Francis's entourage will be similar, if smaller. We strongly advise that you observe them at the meeting point—the Vale of Arden—rather than at the French camp or on the road. You will see 200 mounted archers of the royal Guard dressed in golden coats heading the cavalcade, followed by 200 gentlemen of the royal household, and a hundred Swiss Guards on foot. Twelve trumpeters then precede the major nobles of France: the DUC DE BOURBON, ADMIRAL BONNIVET, and KING FRANCIS himself. The man next to him holding the French sword of state in a sheath of blue velvet covered with golden fleurs-de-lis is GALEAZZO DA SAN SEVERINO, Master of Horse. Behind them will come the DUKES OF LORRAINE, ALENÇON, AND VENDÔME, all the Cardinals of France, and the Knights of the Order of Saint Michel.

As the English caravan leaves its camp it will suddenly be joined by an enormous phalanx of INFANTRY, including the entirety of the ROYAL GUARD all decked out in Tudor livery adorned with golden roses. They will take up positions at the front and back of the entourage as well as forming up on the king's flanks in a parade now 4,000 strong. You may well note a certain amount of toing and froing as scouts are sent off to check the progress of the French, while French scouts scurry back to report to their masters. On at least one occasion the whole troupe will come to a halt as the king and his counsellors discuss whether it is safe to proceed. However, by around seven o'clock you should be entering into the VALE OF ANDREN, where the two entourages will take up positions opposite each other on raised earthen mounds.

The music will now fall silent as the two sides come to a silent halt. Look for King Francis to make the first move, shuffling his horse forward a couple of steps, followed by his three main attendants. Henry with Sir Henry Guildford, the Earl of Worcester, Sir Richard Wingate (ambassador to the French court), and of course CARDINAL WOLSEY. Note, Wingate is wearing a splendid brocade cloak given to him by Francis. It is, as you will see, no match for the KING'S OWN OUTFITS. Henry will be dressed in a silver doublet and cloak slashed with gold and hung with jewels. Around his neck you will see the Order of the Garter and a huge Saint George pendant, all topped by a black hat and plume. Francis is dressed in the currently fashionable latticework jerkin, called a *chammer*, combining gold, silver, and gems, set off by white leather riding boots and, like Henry, a black bejewelled velvet cap.

A sudden blast of the sackbuts will see both kings' companions halt as the monarchs ride on to the appointed meeting place—a spear stuck in the ground. Both entourages will cheer lustily as the kings doff their hats and then, still on horseback, embrace. They will suddenly be joined by running footmen who will accompany them arm in arm as they disappear into a small tent. While the kings are within, you will notice that the English entourage does not break ranks and remains on the mound. The French, however, will become curious and drift over to the English side for slightly awkward conversations. After about an hour the kings will reappear and there will be a great deal of hugging and embracing of each other's companions before a worried-looking Wolsey calls a halt to proceedings and both sides return to camp before sunset.

FRIDAY, JUNE 9TH: THE TREE OF HONOUR

The TOURNAMENT proper will begin with the hanging of shields on the TREE OF HONOUR. Located on the edge of the tournament field, the tree is made from entwined branches of hybrid raspberry and hawthorn, symbolising the new links between France and England. The whole thing is fixed to a stone column and surrounded by wooden railings. At about half past nine, the two kings, sixty noblemen, and sixty guards, drawn from both camps, will be meeting at the tree to HANG SHIELDS.

Pride of place, acknowledging that the field is on English soil, will be going to Francis, whose shield is the first to be hung; Henry's will follow to his right. The shields of fourteen nobles who issue the tournament challenge to all comers will be added below. Finally, three shields will be added to the lower branches to signify the three components of the tournament to come: a GREY AND BLACK SHIELD for jousting, GOLD AND TAWNY for the tourney, and SILVER for foot combat. Over the next couple of hours, the CHALLENGERS IN THE TOURNAMENT will come to the Tree, touch the three shields and place their own shields on the railings. Eventually there will be over 200 challengers and their coats of arms.

SUNDAY, JUNE 11TH: BRING ON THE BANQUETS

At some point in mid-afternoon the sound of CANNON will signal Henry's departure for the French camp and Francis setting out for the English camp. They will meet briefly at the field before being received by each other's QUEENS. If you are able to slip into the BANQUET you will find that the monarchs have their own dining zone (not to be entered), while the rest of the hall has been divided into two by a hanging wall of tapestries.

On one side, over a hundred LADIES OF THE ENGLISH COURT will dine, attended by twenty gentlemen who hover but do not eat. On the other side the leading French nobles DUC DE BOURBON and ADMIRAL BONNIVET dine with their entourages. After dinner there will be DANCING to music from tabors, pipes, and viols, with a guest spot for Francis's own band of FIFES AND TROMBONES, who will be laying down a dance "in the Italian style." The key feature of this dance is that, unlike most court masques, the dancers call upon members of the audience to join them.

MONDAY, JUNE 12TH: JOUSTING BEGINS

On Monday morning much of the English camp will be making its way to the FIELD OF THE CLOTH OF GOLD (as it is not yet known). Located just inside the Pale de Calais, midway between Guines and Ardres, the field is 900 feet long and 300 feet wide, enclosed by an eight-foot ditch, which in turn has created a high earthen bulwark. Within, the field is marked by wooden railing

HENRY DISPLAYS THE CROWN JEWELS, IN A PAINTING AFTER HANS HOLBEIN. BEFORE THIS DEPICTION, STANDING WITH LEGS APART WAS CONSIDERED IMPOLITE.

and posts. Both ends of the field are entered through temporary TRIUMPHAL ARCHES and at the Guines end this is flanked by two sturdy ARMING CHAMBERS for the two monarchs. The Tree of Honour can be seen at the Ardres end just beyond the arch. On the left of the field is the just-completed QUEEN'S PAVILION, from which the royal Households and senior peerage will be watching the day's events; on its right is a somewhat rickety three-storey pavilion for everyone else. Henry has asked for a further deep ditch to be dug in front of it to prevent incursions from the stands, but the soft earth and likelihood of rain mean that this would imperil the building's rather shaky foundations. Instead an additional line of railings has been introduced.

In the very centre of the field are the cloth-draped wooden walls that make up THE TILT. Only introduced into jousting in the late fifteenth century, this allows much greater control and precision on the part of the knights, leading to fewer injuries. A second set of railings on either side of the tilt helps to keep the horse running straight.

Around noon, the two queens—CLAUDE AND CATHERINE—will make their entrance, swiftly followed by Henry and Francis leading their teams of TENANS—the issuers of the chivalric challenge on which the tournament is based. They will be followed by two teams of VENANS, or challengers, led by DUC D'ALENÇON and ADMIRAL BONNIVET. There will then follow a great deal of presentation, ceremony, doffing of lances, and bowing. Finally THE JOUSTS will begin.

TUESDAY AND SATURDAY, JUNE 13TH/17TH: RAINY DAYS AND WRESTLING

The next five days of the tournament are going to be interrupted by squalls and sometimes heavy summer rain showers. TUESDAY THE 13TH will be rained off altogether and the field will become so sodden that the organisers will have to remove the counter-lists from the tilt; this will lead to much poor-quality jousting as

THE TOURNAMENT FIELD AND TREE OF HONOUR—YOU'LL BE SEATED ON THE LEFT.

riders are unable to keep their charges consistently close to the centre. WEDNESDAY THE 14TH will be equally dismal, but enlivened in the afternoon by a large WRESTLING COMPETITION in the mud between English guards and Breton wrestlers, as well as an ENGLISH ARCHERY DISPLAY.

Things will be dry enough on THURSDAY THE 15TH for JOUSTING to recommence, with both Henry and Francis putting in an appearance. HENRY'S JOUST WITH MARSHALL LESCYN is the sporting highlight of the day and a sartorial triumph, with the king's cloak decorated with lozenges and eglantine flowers of gold. The following day it will rain again and, though a few jousts will be held, neither kings nor queens, nor anyone very special, will show up. However, on SATURDAY THE 17TH the sun will return and a decent day's jousting will commence. Both Henry and Francis will appear to be in particularly good fettle, breaking eighteen and fourteen lances, respectively, in their contest with a team led by the Earl of Devonshire.

SUNDAY, JUNE 18TH: BACK TO THE BANQUET

The camp will be thick with rumour and gossip today, as FRANCIS will be paying a surprise early morning visit to HENRY'S APARTMENTS, after which they will go to MASS together. Henry will then depart for Ardres and Francis will be relaxing in Queen Cath-

GO EASY ON THE ROAST SWAN TO AVOID THE FATE OF THE POOR UNFORTUNATE THROWING UP ON THE LEFT.

erine's apartment before DINNER. Again the dining hall will be divided during the meal, with men and women on different sides, but afterwards the tapestries will be cleared and the DANCING will begin, led this time by Francis himself. He will be entering the hall with ten companions in long, velvet-hooded gowns adorned with plumes.

MONDAY AND TUESDAY, JUNE 19TH/20TH: LAST JOUSTS, COSTUMES, AND THE TOURNEY

The final days of the JOUST will be clear and dry. On the Monday the two kings will be present and will conduct an elaborate exchange of horses and gifts. On Tuesday they will return to the tilts.

Do try to look out for KING FRANCIS'S CLOTHING, which, on the days he jousts, has been designed to signal through a variety of symbols—brooches and embroidery—carefully calibrated chivalric phrases. Members of the French court will be on hand to help decode these, so feel free to ask. Tuesday's motto, for example, is

"Heart fastened in pain endless/when she/delivereth me not of bonds."

Tuesday will also be the day of the TOURNEY, in which combatants fight on horseback and in pairs, but in open space rather than across a barrier. Weapons include blunted swords, staves, and clubs but much of the skill and excitement is in the quality of the HORSEMANSHIP as riders attempt tight turns and precise moments of ACCELERATION to ensure their blows have maximum force. The very best competitors will time their runs and blows to occur in front of the queen's pavilion.

WEDNESDAY AND THURSDAY, JUNE 21ST/ 22ND: COMBAT ON FOOT

The FOOT COMBAT of these two days involves individual bouts fought across a central wooden barrier—thus preventing grappling and wrestling and keeping the emphasis on weapon skills. Combatants will use short swords, spears, and pikes. It is particularly worth noting the quality of HENRY'S ARMOUR. Although unable to use his most modern and sophisticated armour, after lengthy and complex negotiations with the French, Henry is able to wear a suit that displays many technical innovations of the royal armoury. Francis has insisted on the use of closed visors and the *tonlet*—a protective skirt of metal plates. To this Henry will have added *cuisses*—laminated steel breeches—and *lames*, which are the articulated steel plates bound to leather strips that provide unrivalled protection and freedom of movement at the joints.

FRIDAY, JUNE 23RD: MASS AT THE TILTYARD

Overnight, a huge workforce has been working furiously to repair the TILTYARD and build a vast public stage over it. On this they will have built a large chapel richly furnished with tapestries and bejewelled crucifixes (you may recognise them from the royal Chapel back at the English camp). Amazingly, the task will have been accomplished. You will find the CHAPEL positioned between the queen's pavilion and the general viewing galleries.

MASS will begin at noon, with Cardinal Wolsey leading proceedings. Note the relative height of the various clerics on the platform: Wolsey is positioned just above Cardinal de Boisy; both are higher than France's other cardinals, while the bishops of both nations are lower still. Music and singing come from the ROYAL CHOIRS of both nations and the organ from Henry's chapel. The French organist PIERRE MOUTON, accompanied by voice, sackbuts, and coronets, will play a particularly good rendition of the liturgical prayer *Kyrie*.

Look out for CARDINAL BOURBON, who will take the gospel to both kings and both queens for them to kiss. Henry and Francis will also embrace with the PAX—a holy kiss—while the queens join arms. Just prior to the Elevation of the Host, the point at which the wafers and wine become the body of Christ, be sure to look up. You will see a DRAGON appear in the sky. As one contemporary would later report, "Lo! Flying in great loops, a slender and hollow monster stretched out in the sky." While later paintings will depict a great fire-spitting beast, you will see that this is a large, if sturdy and highly decorated, FRENCH KITE. Cardinal Wolsey's face, upstaged by this event, is worth your consideration.

SATURDAY, JUNE 24TH: THE LAST SUPPER

Once again the two kings will EXCHANGE CAMPS AND PALACES for a final FAREWELL BANQUET. On this occasion both will have dressed in their MASQUE OUTFITS before the meal. If you are alert you may be able to see Henry ride out dressed as Hercules in a woven gold lion's pelt carrying a great wooden club wrapped in green damask. His company are dressed as a mixture of HEBREW KINGS and Christian warriors like KING ARTHUR and CHARLEMAGNE. Tonight's feast will be concluding with the presentation of jewels, gems, and other prizes to the winners of the tournament. Drink up, but do return to the Fortescue pavilion for your DEPARTURE.

The St Louis World's Fair

AUGUST 27–SEPTEMBER 3, 1904 ✵

ST LOUIS, MISSOURI, US

"MEET ME IN ST LOUIS, LOUIS. MEET ME AT the Fair." In 1904 there was only one place to be and one place to go: the greatest show on earth, the maddest cabinet of curiosities ever assembled, the Louisiana Centennial Purchase Exposition, aka the 1904 WORLD'S FAIR. Join the throng—almost twenty million all told—who wandered through its palaces and pavilions. Be awed by the new technologies of electric generators, automobiles, and X-rays at the Fair, and then head over to its wackier cousin, THE PIKE, for a riot of candyfloss and beer, themed cafes, merry-go-rounds, and water chutes. You could almost be in Disneyland. Go back to the past and say, "Welcome to the Future."

And that's not all, folks. The Fair coincides with the THIRD MODERN OLYMPICS, which were due to be held in Chicago but have been artfully co-opted by Louisiana. So your trip includes top sporting action and an insight into the Olympics as it once was—accessible, amateur and just a little bizarre.

BRIEFING: REVENUE ACTS AND MASSACRE

The St Louis World's Fair or, to give it its full title, the LOUISIANA CENTENNIAL PURCHASE EXPOSITION, opens on April 30, 1904. It is actually a year late to be a true centennial—commemorating the deal stuck by the US Federal Government with the French state to acquire about a third of the continent. But that is understandable in such an immense undertaking. Before any construction began, a workforce of over 10,000 labourers had to clear a 1,200-acre site on the western edge of St Louis, near the campus of Washington University, dynamiting innumerable tree stumps, draining swamps, diverting streams, and building and levelling hills.

On your arrival, more than 1,500 buildings will have been erected in the fairgrounds, connected by some 75 miles of roads, railways, and sculpted walkways. At the heart of it are great white beaux-arts pop-up palaces, warehouses of industry, invention and commerce, dressed in stucco towers and tracery.

The driving force behind the fair is the ex-mayor of St Louis, DAVID ROWLAND FRANCIS, who has orchestrated its financing, supervised its building, and gathered many of the Exposition's exhibits. Like all of the world's fairs of the belle époque, there are many agendas at work: boosting St Louis on the world stage; celebrating America's new imperial and global power; showcasing commercial products and technological innovations, and, if even a fraction of the fifty million who paid to go to the Exposition Universelle in Paris in 1900 come through the doors, then turning a buck for all the private investors who ponied up to put this thing on.

Devotion to the dollar and fear of losing box office is the reason that the Fair is also hosting the 1904 OLYMPIC GAMES. Originally awarded to the patrician athletes of Chicago by the International Olympic Committee, the St Louis people immediately threatened to hold their own Games if that plan went ahead and muscled Chicago out of the picture.

It should be noted that, despite the apparent universalism of the World's Fair and its vision of humanity, the usual RACIST SEGREGATIONIST rules of early twentieth-century America apply, with separate but distinctly unequal facilities for black and white visitors. Thus many African American leaders have protested prior to the Fair and called for a boycott.

Time travelers who feel that their skin tone or ethnic heritage might cause them problems when dealing with this particularly unpleasant aspect of gilded-age America should not hesitate to contact the WAG management, who will endeavour to make suitable arrangements to ensure a racism- and hassle-free experience.

THE TRIP

You will arrive on the morning of Saturday, August 27 outside of the INSIDE INN, "the only hotel within the grounds of the World's Fair" as its brochure proclaims. Situated in the southwest corner of the park, the Inn is the giant pop-up creation of noted hotelier E.M. Statler, a colonial revival concoction with 2,257 rooms and 2,000 staff. You are booked for the week on the *American Plan* (full board) at $3 day. This includes free entrance to THE FAIR, which is open daily except Sundays from 9am to 10pm. THE PIKE, the Fair's more raucous sibling, stays open until midnight. You will of course be able to access your hotel and the fairgrounds, which are fabulously lit at night, at all times.

MEET ME IN ST LOUIS, LOUIS. THIS POSTCARD SHOWS THE PALACE OF MACHINERY, WITH ONE OF THE FAIR'S FAMED DIRIGIBLES ALOFT.

PHARUS-MAP
WORLD'S FAIR ST. LOUIS
1904.
LIBRARY of CONGRESS
Two Copies Received
JUN 1 1904
Agriculture
Horticulture
Live Stock
Camping Ground
Philippine Exhibit
Nursery
Physical Culture
Forestry
Transportation
Jerusalem
Machinery
Fine Arts
Grand Basin
Varied Industries
Electricity
Education
Manufactures
Mines and Metallurgy
Liberal Arts
Sunken Garden
Plaza
Tyrolean Alps
Irish Village
Asia
THE
Cairo
PIKE
Battle Abbey
Japanese Village
Skinker Road
Forsyth Avenue
Pennsylvania Ave
Concourse Drive
Lindell Boul.
Oakland Ave.
INTRAMURAL RY.
INTRAMURAL RAILWAY
Hotel Jefferson Inn

For travelers in search of a more populist experience, we offer alternative accommodation at CAMP LEWIS, an 85-acre campground created by eccentric magazine publisher Edward Gardner Lewis just beyond the eastern perimeter of the Fair. At the camp—a square of tent cabins with wood floors, iron beds and electric lights—there are nightly campfire sing-alongs and musical entertainments. Do note, however, that alcohol is banned. At the north end of the camp you will see an octagonal tower, the headquarters of Lewis's publishing empire, which will sport the world's largest carbon arc searchlight. Lewis himself arrived in St Louis selling insecticides and snake oil, before buying a local periodical that he turned into *Women's Magazine*, the nation's bestselling title.

EXPLORING THE FAIR

When you are ready to leave the Inside Inn, by all means plunge straight into the Fair. Outside the hotel you will find yourself at the southern end of the PLATEAU OF STATES, with the UTAH pavilion in front of you and INDIANA to your left. However, we suggest you hop on the MINIATURE RAILWAY to the Fair's main entrance (STATION 17), in order to take in its sheer enormity. To your right you will see a recreation of the TYROLEAN ALPS, complete with mountain villages and snowcapped peaks. Behind that you might glimpse the peat smoke curling up from the equally large IRISH VILLAGE. These are just two of the exhibits on the THE PIKE, an area that lies outside the perimeter wall of the Fair but is still part of the show: a mile-long strip of the weirdest and most wondrous attractions, rides, recreations, and spectaculars that the early twentieth century can offer.

Resist the urge to explore them now, however, and head through the entrance, where you will find yourself in the PLAZA OF ST LOUIS looking towards the immense, ornate GRAND BASIN. Arranged around you are vast PALACES devoted to education, economy, industry, agriculture, horticulture, and arts, while to the southwest is the PLATEAU OF STATES, and to the northeast the INTERNATIONAL PAVILIONS and the phenomenally popular

Aeronautical Concourse. Right at the edge of the grounds is the university's gymnasium and its brick stadium, Francis Field, site of most of the events in the Olympic Games.

The intramural railway is your best way to get around the Fair. It encircles the whole site with seventeen stations along the way. There are also miniature railway lines: one runs the length of the Pike on Administration Avenue; the second goes from the Pike to the Boer War Concession and Philippine Village. Other modes of transport are available for crossing the Grand Basin and negotiating the various lagoons and canals. Moored on the jetties around the edge of the basin you will find launches with musicians and dancers, Venetian gondolas with singing gondoliers, swan boats, south sea outriggers, Hawaiian surf-boats, and Australian catamarans.

Two landmarks that you should be able to see from almost anywhere in the Fair are the DeForest Wireless Telegraph Tower and the Observation Wheel, both located right at the heart of the site. The Tower features an electric elevator that whisks you up 100 meters above the Fairgrounds. The wheel is the same one that dominated the skyline of the 1893 Chicago World's Fair, designed and built by the late George Ferris, and has been dismantled and reassembled here. Listen out for the giant air brake used for stopping the thing.

THE PALACES

The Palace of Fine Arts, one of the few buildings that will outlast the Fair, overlooks the Grand Basin. It is an art exhibition to rival the size of any modern biennale, with 135 galleries of American and world art, arranged by nation, and a dome larger than St Peter's in Rome. Much of the art, however, is numbingly mediocre, though you might seek out new works by Whistler, Sargent, and Rossetti. Lower down the hill and directly in front of the palace is the Festival Hall, the Fair's main auditorium, where you can marvel at the world's largest organ, equipped with 10,059 pipes. There are daily recitals at 10 cents a go, while orchestral performances in the early evenings are 25 cents.

Beyond here the PALACES OF MANUFACTURES and VARIED INDUSTRIES feature a riot of consumer goods—textiles, pottery, glass, lace, needlework, clothing, cutlery, furniture, watches, jewellery, and clocks—and showcase working factories making shoes, pens, paper boxes, and hats. Nearby is the PALACE OF MINES AND METALLURGY, with its towering twin obelisks. Inside are cabinets and cabinets of gems, ores and metals, and mining equipment, and, rising above them, a mammoth cast-iron statue of the Roman god VULCAN holding a newly forged spear; this was commissioned by Birmingham, Alabama, in homage to its iron and steel industries. Outside in the MINING GULCH is an underground mock-up of a mine with its own themed restaurant.

It is 1904 and electricity is still little short of miraculous. Take a moment to look up at the PALACE OF ELECTRICITY, and you will see the nude figure of Light holding aloft a star, and at her feet the crouching curs of Darkness. Fabulously illuminated at night, the palace is stuffed with revolutionary electrical equipment and appliances: X-ray machines, telegraph booths, telephones, and batteries. The newly invented FINSEN LIGHT for treating Lupus will be drawing big crowds.

The PALACE OF MACHINERY, distinguished by the square towers at each of its corners, has a joint function as the Expo's power station. You will find many visitors simply awed and overwhelmed by the scale and brute metallic sculpture of its generators, dynamos, and heavy industrial metal presses. Further along is the PALACE OF TRANSPORTATION, whose three giant archways give it the feel of a great urban rail terminus. Inside, the past has been preserved as stuffed mules, dog-sled teams, oxen, and the horses of horse-drawn carriages. But the future has arrived in the shape of electric trolleybuses, motorboats, and automobiles. For the moment, though, rail is king, as illustrated by a monster 162-ton locomotive and tender mounted on a revolving turntable.

THE WEST LAGOON—ON A STATE-OF-THE-ART STEREOSCOPIC CARD.

The PALACE OF LIBERAL ARTS is a mixed bag, with exhibits featuring photographs and photographic equipment, business machines, typewriters, medicines, coins, and a lot of medals. It also houses New York State's 18-foot-high salt carving of a lighthouse and the Chinese delegation's huge array of ancient books, medieval armour, and weapons. Travelers are asked to resist the temptation to use the Photoscope Company's self-service photo booths (six for a quarter).

The Midwest is the Midwest and so the PALACE OF AGRICULTURE is the biggest building on the grounds, indeed the biggest at any World's Fair ever, housing 1,000 exhibitors, and enclosed by a garden planted with one million roses. Inside you will find a working dairy farm, cider and rice mills, and the latest in humane animal slaughter. The US state delegations have really pulled their finger out, too, with butter sculptures often their centrepiece; they include Minnesota's 20-foot likeness of Father Hennepin discovering the St Anthony Falls, and North Dakota's life-sized figure of President Theodore Roosevelt on horseback. Louisiana has a tableau of Mephistopheles made from sulphur, Lot's wife carved from a block of rock salt, and Miss Louisiana crafted from a giant sugar cube. Kansas is showing off a nine-foot Indian made out of wheat. Best of all, perhaps, is the home state's 2,900-pound cream cheese sculpture of a Missouri maid milking a cow. Such themes—curously addictive—continue in the PALACE OF HORTICULTURE, a cornucopia of fruits and nuts. Mississippi has installed a horse made entirely out of pecans; California, an elephant made out of almonds and a grizzly bear constructed from prunes.

The PALACE OF FORESTRY, FISHERY AND GAME, to the north, is very heavy on taxidermy, with slightly dreary stuffed animals from every US state, but the aquarium section, with its 60 large tanks and saltwater basin, is more fun. And we must direct travelers to the fabulous floral displays outside: a FLORAL CLOCK over 20m across and made up of 10,000 plants, its moving hands coated in scintillating pick blooms; and an immense FLORAL MAP OF THE UNITED STATES with each territory picked out in a characteristic plant—alfalfa in Kansas, blue grass in Kentucky, wheat in North Dakota.

A STRANGE MISCELLANY

Nearby the palaces are a miscellany of attractions worth taking in. A good first stop is GRANT'S CABIN, a ramshackle country bungalow built by General Ulysses S. Grant in 1854, when long before his Civil War call-up he had resigned from the Army. The building has been brought here at the behest of Cyrus F. Blanke, the St Louis businessman and coffee-roasting magnate, and doubles up as a COFFEEHOUSE.

From here you might wander over to the vast PHILIPPINE VILLAGE, where the spoils of the new US Empire are on display. The Philippines were acquired only in 1898, after the Spanish-American War, and the US Department of War has assembled 1,200 Filipinos to perform everyday life in a giant "reservation."

Equally extraordinary is the JERUSALEM EXHIBIT, a 10-acre replica of the old city that includes the Wailing Wall, Dome of the Rock, and Church of the Holy Sepulchre, which is populated by a cast of Jews, Muslims, and Christians. Join the faithful and "follow in the steps of Jesus," or take a ride on a donkey.

And, for war re-enactment fans, there is the BOER WAR EXHIBIT. Right at the heart of the fairground, this huge arena will stage twice-daily battles from the recently concluded Anglo-South African war. The shows take up to three hours and feature 600 actors, many of them veteran combatants (from both sides)

DANCERS PERFORM AT THE PHILIPPINES VILLAGE: ONE COMPONENT OF WHAT WAS ESSENTIALLY THE FAIR'S ANTHROPOLICAL ZOO.

of the real war. Do make sure you stay for the climax, when the Boer general Christiaan de Wet escapes on horseback by leaping from from a cliff into a pool of water. The concession also features a British Army encampment, Zulu, Swazi, and Ndebele villages, and stages parades, sporting events and horse races.

SPECIAL DAYS AT THE FAIR

Except for Sundays, when the park is closed, every day in St Louis has some special attraction—a spectacle, conference or parade. Our visitors will be arriving too late to witness the National Fireman's Tournament, or the pleasures of Opticians Day. However, there will be plenty of unusual action to engage you.

SATURDAY, AUGUST 27. Today will be Liberal Arts Day and an excuse for a giant parade of automobiles dressed in carnations, roses, and ribbons. Follow the crowds, too, to the aeronautical concourse for the Fair's second GREAT BALLOON RACE.

SUNDAY, AUGUST 28. The park will be closed, but visitors may like to make their way to the old horse-racing track, where America's first speed freaks will be racing their cars in pursuit of the LOUISIANA PURCHASE TROPHY. Do take care not to get too close to the action, as celebrity racer Barney Oldfield will smash his Green Dragon car into a tree, killing two unfortunate spectators.

MONDAY, AUGUST 29. The Opening Ceremony of the 1904 ST LOUIS OLYMPICS in Francis Field Stadium.

TUESDAY, AUGUST 30. The Olympic Marathon. Indiana Day begins.

WEDNESDAY, AUGUST 31. Mining Gulch Day sees 10,000 free water-melons given out around the Old Anthracite Mine. You may well also notice a crowd gathering around the observation wheel in the after-noon where two couples will be married in the top gondola. It's Indiana Day again, too.

THURSDAY, SEPTEMBER 1. It's still Indiana Day, and Tennessee Day, and the first of a two-day horse show at the livestock forum.

FRIDAY, SEPTEMBER 2. Last chance to celebrate Indiana Day.

SATURDAY, SEPTEMBER 3. A morning march and performance from LA GARDE RÉPUBLICAINE, the French Army's finest band. Catch the last of the OLYMPIC TRACK AND FIELD events and the finals of the TENNIS tournament. This is also going to be Commercial Vehicle Day with a parade of over 300 automobiles of various kinds and, to top it all off, there will be a parade of the Kansas-based masonic fraternity, the SONS AND DAUGHTERS OF JUSTICE.

US GOVERNMENT AND STATE PAVILIONS

The US government's chief contribution to the Fair is a Romanesque-columned building on "Government Hill." Every department of the federal government is putting on a show here and most, as you might imagine, are pretty dull. However, the PLATEAU OF STATES around the hill is more inviting, with its eclectic mix of period reproduction houses, homespun kitsch, and takes on Roman classicism and colonial revivals. The TEXAS PAVILION is a balconied colonial mansion pulled into the shape of the state's five-pointed star; VIRGINIA has gone with a reproduction of the MONTICELLO home of President Jefferson. MISSISSIPPI has built a copy of BEAUVOIR, the southern mansion bequeathed to Jefferson Davis where the Confederate president spent the last years of his life; its display of the clothes he was purportedly dressed in when captured by the Union is an attempt to scotch the rumour that he was dressed in a women's coat. OREGON has built a reproduction of FORT CLATSOP LEWIS and CLARK'S 1805–6 STOCKADE at the mouth of the Columbia River. It has a big restaurant worth trying, while beneath the vast rotunda of the PENNSYLVANIA building there is a busy smoking room.

One interesting interloper, and worth a visit for the name alone, is the HOUSE OF HOO HOO, built by the International Order of Hoo Hoo—a transcontinental masonic lodge for executives, journalists and bankers in the global lumber industry. Visitors will be seeing the second version of the pavilion. The first burned down in June, requiring the Order to build this replacement.

THE INTERNATIONAL PAVILIONS

The ten blocks or so on the northern edge of the Fair, to the east of the Palace of Transportation, are home to the INTERNATIONAL PAVILIONS. They range from the modest two-storey houses of GUATEMALA and NICARAGUA or SWEDEN's old coach house, to BRAZIL's grandiose dome-topped mansion, an INDIAN pavilion modelled on an Agra palace, and CHINA's replica of the summer palace of Prince Pu Lun. The BRITISH house is a copy of the red-brick Orangery in Kensington Palace in London, with a rather fine billiard room that, alas, you can't play on. The FRENCH are also thinking big with their replica of the Grand Trianon palace at Versailles, though exhibits here are mundane, with more ornate gold-leaf furniture than you can shake a stick at. CANADA's modest wooden offering is full of maps and promises of land for settlers. Your best bet for something approaching conviviality is to head for the Brazilian pavilion, where free coffee is available all day and each evening there's one of the Fair's top parties (an invite to this is included with our travelers' tickets).

FOOD AND DRINK AT THE FAIR

Your "American" deal at the Inside Inn allows you to eat there three times a day. However, we would encourage travelers to sample some of the hundreds of restaurants, cafes, kitchens, and coffeehouses that the Expo offers. The Fair's hidden secrets are the BARBECUE CONCESSIONS serving up hot roast-beef sandwiches for just a dime and fulsome plate lunches at 20 cents a time. Find them on the Pike next to the Japanese Tea Garden and New York to North Pole, inside the Philippine Village, outside the Palace of Liberal Arts, and in front of the US government building.

Foodies will want to head for the MODEL PAVILIONS in the Model City, whose kitchens are being organised by celebrity cook SARAH TYSON RORER, the author of no fewer than 75 cookbooks and booklets including, of course, *The World's Fair Cookbook*. She is the first healthy-eating guru in America, cautioning: "a large quantity of fried foods may be eaten without nourishing the body; and of one thing we are quite sure, they always tax the digestive organs." Expect plenty of broth and boiled vegetables.

For those who like themed dining, look no further than the ANTHRACITE COAL MINE RESTAURANT. The waiting staff are dressed as coal miners, suitably blacked up with coal dust, and will take you to a table in a subterranean simulated mine. Tables are lit by miners' lamps, while an actual mine fan is pumping fresh air into the dining room. The food is mainly hearty Dutch fare.

For a sampling of what the average Joe might chose, try BEAUD'S RESTAURANT in the Palace of Manufactures. A typical lunch might include baked whitefish in a butter sauce, baked chicken pie, roast-beef rib, or a Waldorf salad, with an ice-cream sundae for dessert. In the state pavilions the LOUISIANA AND TEXAS RICE KITCHEN will be offering baked red snapper with rice, as well as rice breads, muffins, rice lemon pie, and rice ice cream.

PHILIPPINE CUISINE can be sampled at Cafe Luzon, Cafe Michel, and the Nipa Barracks Cafe, inside the Philippine Village.

There's much grazing to be done, too. At the Palaces of Agriculture and Horticultural FREE SAMPLES are dished out, and many of the main commercial exhibitors are lavish with tasting dishes. There is also an inexhaustible supply of baked beans, bread and butter at the MINNESOTA state pavilion.

THE AERONAUTICAL CONCOURSE

Your time at the St Louis World's Fair comes just nine months after the Wright brothers flew their historic twelve-second self-propelled flight...and America has gone plane-crazy. A large field on the northern edge of the Fair, behind the International Pavilions, has been laid out as the AERONAUTICAL CONCOURSE and equipped with a vast hangar, two gas plants for blowing up balloons, and, somewhat quixotically, a 30-foot high fence to keep out the wind. The US government has offered a prize of $100,000 to anyone who can fly a one-seater self-propelled plane around a one-mile triangle, maintaining a speed of at least 15 miles per hour.

Depending on which day you visit the Concourse, you will able to see a selection of gliders, kites, hot-air balloons, motorised balloons, or dirigibles, and a variety of experimental airplanes, take to the air...or not. None, unfortunately, will be doing well enough to claim the federal government prize, and most of the machines that do get airborne have enormous problems controlling their direction of flight. Two splendid machines to look out for are the VILLE-DE-SAINT-MANDÉ and the CALIFORNIA ARROW, huge hydrogen-filled balloons from whose perilous cradles and motors are attached. The former is the creation of French aeronautical ace Hoppolyte François; the latter has been built by America's leading

THE CALIFORNIA ARROW AIRSHIP, WITH AERONAUT ROY KNABENSHUE, IN ITS FIRST SUCCESSFUL FLIGHT AT THE FAIR.

balloonist, Thomas Scott Baldwin. Visitors are reminded to be cautious about smoking in their vicinity.

THE PIKE

World's Fairs have been going for the best part of a century and, over recent decades, have been attracting an unruly fringe of circus acts, zoos, and fairground rides. At the 1893 Chicago World's Fair they were formalised on a strip called the Midway Pleasance, and at St Louis they are concentrated along the huge paved boulevard known as THE PIKE. Accessible without a fairground ticket, open till late and with alcohol liberally available, this is the more raucous end of the Fair: keep your eyes open for pickpockets and hard-luck stories. But it's a lot of fun and, with more than forty shows of one kind or another, it's impossible to do it justice in one visit. Take your time and explore it block by block.

WEST END: SUBS AND A ZOO

Your trip down the Pike starts with a dose of ethno-regional kitsch with a tram ride offering stops in the TYROLEAN ALPS, an IRISH VILLAGE complete with replica Blarney Stone, and a SPANISH STREET fronted by buildings from Madrid, Barcelona, and Andalucía. Each have their strong suits: the strudel and coffee are good in the Tyrol; the Irish Village features imported sod; and fans of palmistry will find Spanish gypsy card reading and fortune telling. More outdoor action is available at HUNTING IN THE OZARKS, where you can take a train ride through northern California and shoot real game from the comfort of your seat.

And the fun goes on. OVER AND UNDER THE SEAS puts you on a great tilting platform and carries you on a submarine journey across the Atlantic that surfaces on the Seine in Paris, by way of great tableaux of marine life and shipwrecks. The return journey is via airship and takes in a violent lightning storm—you are advised to keep your coat on.

The end of the West End block is entirely taken up by HAGENBECK'S ZOOLOGICAL AND TRAINED ANIMAL CIRCUS, where a 3,000-seat arena hosts performances by seals, snakes, horses, monkeys, lions, and tigers, and (most popular of all) baby elephants sliding down water chutes. Rides on llamas, ostriches, and zebras are available.

SOUTH SIDE: ASIA, PARIS, AND BABES

MYSTERIOUS ASIA is a huge mashup of predominantly British imperial possessions. There are enormous models of the Taj Mahal, the Golden Temple in Rangoon, downtown Tehran, and parts of Ceylon and Delhi. You should catch PRINCESS RAJAH'S

THE TYROLEAN ALPS—A NICE SPOT FOR LUNCH.

HOOCHIE KOOCHIE DANCE here—a sensation at the Fair, it combines suggestive belly dancing with castanets and chair-balancing tricks. Elephants and camel rides are on offer, too.

The huge PARIS ZONE is amongst the best places to party on the Pike. The champagne is likely to be flowing at CAFÉ CHANTANT, with its rotating cast of acrobats, drag queens, and comedians, and there is open house poetry at CABARET BRUANT.

One fabulously bizarre element of this block is the INCUBATOR EXHIBIT. In the midst of The Pike the organisers have thought it a good idea to set up what they consider to be a state-of-the-art hospital ward with the newly invented baby incubators not merely on show but in action, with a team of nurses and doctors looking after their unfortunate charges. By the time our visitors arrive, the disastrous state of hygiene in the exhibition has been remedied, with the insulation of glass walls between visitors and babies.

Two shows towards the end of the block can be skirted. OLD PLANTATION is an unspeakably saccharine account of the Antebellum South and the "olden life of the American negro," while BATTLE ABBEY, though a splendidly faux-medieval castle with drawbridge and crenellated towers, is little more than a venue for cycloramas of famous American battles. Better to spend your time with BEAUTIFUL JIM KEY the equine sensation—a horse who can spell, do math, sort mail, tell the time and quote the Bible.

EAST SIDE: FLOODS, BATTLES, AND FIRES

Things take a decidedly watery, indeed stormy, turn at the EAST SIDE of the Pike, where visitors will find the GALVESTON FLOOD, BOYNTON'S NAVAL EXHIBIT, and HALE'S FIREFIGHTERS. The first is an epic re-creation of the devastating flood of 1900 that hit the Texas town, wiping out most of its buildings and 6,000 people in a single day. BOYNTON'S is a huge pool of water, about the size of a football pitch, landscaped with cliffs, coasts and islands, on which a huge flotilla of motorised model ships will be re-creating a selection of the best naval battles of the American Republic. The Battle of Santiago Bay (1898), when the Spanish fleet fled from the Americans, is particularly popular. HALE'S is the baby of George C. Hale, former Fire Chief of Kansas City turned showman. Four times a day his boys will be putting out a fire on a six-storey building that then collapses before your eyes.

NORTH SIDE: SHUTES, TURKS, AND SIBERIA

There are two rides you shouldn't miss on the NORTH SIDE. The first is SHOOT THE CHUTES, which puts you on a boat and sends you down a sharp 100-metre drop alongside the best-looking showgirls and the biggest light display on the Pike. Almost equally fun is the MAGIC WHIRLPOOL—a spiral boat trip around a huge fountain and down brightly lit cascades that then drop down into a submarine channel flanked by tropical gardens.

Those with a taste for anthropological zoos will enjoy the stereotypical depiction of hunter-gatherers at the CLIFF DWELLERS, featuring performing Hopi and Zuni "Indians," and ALASKA, where Inuits will be performing on polar landscapes, fighting with wild dogs, hunting reindeer, panning for Klondike gold, and putting on faux wedding ceremonies.

CONSTANTINOPLE and its TURKISH VILLAGE and OTTOMAN HIPPODROME is essentially a shopping experience, as is the vast CHINESE VILLAGE. There is some respite from the souvenirs and knick-knacks, though, with a Chinese theatre, dragon dancing, fire-eaters, and magicians. And, for a more sedate pause in proceedings, you can take the TRANS-SIBERIAN RAILWAY, a train ride through dioramas of St Petersburg, Moscow, Lake Baikal, and a rather grisly mock-up of the Boxer Rebellion in Manchuria.

NORTHWEST: ALL THE WORKS OF THE LORD

The last—northwest—block of the Pike begins with CREATION. In effect its three boat rides survey the Works of the Good Lord: landscapes, cities, and the Seven Days of Creation itself. Be warned, though: the Creation wasn't just a big bang. This ride goes on for over two hours.

Inevitably, the city chosen is Old St Louis, a reproduction of the city in 1803, when the Louisiana Purchase treaty was signed, and as such a rather unassuming array, with a fort, a stockade and church, and small schools. However, there is fun to be had if you look closer, with a fabulous selection of high-wire bicycle acts at work, as well as fine food at the Edelweiss restaurant.

Rather more lively is the theatrical violence of CUMMINS' WILD WEST. Fredrick T. Cummins has assembled a motley crew of 800, over half of them Native Americans, putting on "Cowboy and Indian battles," with attacks on settlers' cabins and lasso and horse-riding trickery. ANCIENT ROME is in similar vein, featuring a great mock-hippodrome with chariot racing, gladiatorial contests, prize fighting, boxing, and broadsword battles.

After all this excitement you might like to retire to the relative calm of FAIR JAPAN, stroll through its "authentic" street scenes, take a human rickshaw, and meet the geisha girls. Alternatively you might just like a nice cup of tea, available in the delightful JAPANESE TEA GARDENS.

FOOD AND DRINK AT THE PIKE

There is no shortage of food and drink on the Pike, with almost every operation doing some kind of catering, not to mention dozens of street sellers offering snacks, candyfloss, and sodas.

If you want to go large, head for the LUCHOW-FAUST TYROLEAN ALPS RESTAURANT, with views over the entrance plaza and the fairground. It is simply enormous, can cater for 2,500 people inside and the same again on its huge terraces. The main dining room is big enough to house a 100-piece orchestra, which should be playing when you visit. You can eat cheaply here—caviar sandwiches for 40 cents, a filet mignon or a schnitzel for just a dollar—but with more almost 200 dishes to choose from, including an exceptional list of Bavarian specialty desserts and reasonably priced vintage Champagne, why hold back?

For something with a little less cream and a little more spice, drop in on the CHOP SUEY RESTAURANT in the Chinese Village or try a curry at BRITISH INDIA in Mysterious Asia. Or, for simpler fare, head to ABC BEERS INDIAN CONGRESS CAFE attached to Cummins' Wild West Show or the ROAST BEEF SANDWICH CAFE in the Paris exhibit.

THE OLYMPIC GAMES

The sporting programme of the St Louis World Fair actually runs for over six months from May to November, and includes—much to the chagrin of Baron de Coubertin and his insufferably stuffy International Olympic Committee—state athletic championships for Missouri schoolboys, intercollegiate basketball tournaments, days devoted to Gaelic sports, and exhibitions of Turnen and Sokol gymnastics. But it is the Olympic Games that are the centrepiece of proceedings, and your trip coincides with the key events.

Don't expect the trappings of modern Olympics. There are no flags or signage—these won't happen for another couple of decades—and there is not that much ceremony, either. Baron de Coubertin, suffering from one of his periodic fits of pique, has elected to stay away.

MONDAY, AUGUST 29TH

Make your way to FRANCIS FIELD STADIUM, hidden away behind the aeronautical concourse, for the OPENING CEREMONY of the Games and the first day of TRACK AND FIELD EVENTS. You will see David Francis (President of the World's Fair) and James Sullivan (President of the American Athletic Union) take a very cursory stroll down a double file of athletes in the centre of the field. Then a small band will start playing, indicating it's time for everyone to disperse and warm up. So much for the opening ceremony.

Do stay on, though, to see RAY EWRY—known since his gold medal winning performances at the 1900 Paris Olympics as the the human frog—win the STANDING HIGH JUMP. He will be back later in the week to win two more golds in the the standing broad jump and the standing triple jump.

TUESDAY, AUGUST 30TH

Today is devoted to the MARATHON, which will start and finish in Francis Field. Thirty-two runners will begin the race but only fourteen will complete it. The temperatures are in the high nineties, the atmosphere humid and muggy, and all the runners will have to contend with great volumes of dust thrown up by Missouri's hapless, rocky roads. There will be no fresh water available to the runners until 12 miles into the race, at which point the American William Grace will drink so much that he will suffer a near-fatal haemorrhage of the stomach 4 miles further on. Others, erroneously believing that fluid intake during the race is a problem, will subsist on wet sponges and brandy-drenched flannels. Look out for FELIX CARVAJAL, a postman who has made his own way to the Games from Cuba, and, having lost his travel money in a dice game in New Orleans, has arrived penniless in St Louis. He will run in heavy shoes and cut-down trousers and, despite stopping to eat and to chat to spectators, will finish fourth. Two Africans who have been working on the Boer War concession on the Pike—LEN TAU and JAN MASHINAI—will be amongst the other finishers; Tau comes a very creditable ninth and after the race will run an additional mile to get away from an aggressive dog.

The American FRED LORZ will be the first athlete to enter the stadium and pandemonium will reign for a few moments until it becomes apparent that he has made at least part of the way on the back of a truck. He will be disqualified. THOMAS HICKS, next into the stadium, will slow to a walk but manage to get himself over the line. It will later transpire that his coach Charles Lucas has been refusing him water during the race while administering him a mad concoction of strychnine, alcohol, and egg whites—all perfectly legal in 1904.

THOMAS HICKS, HEAVILY DRUGGED, MAKES IT INTO THE STADIUM TO WIN GOLD IN THE 1904 OLYMPIC MARATHON.

WEDNESDAY, AUGUST 31–SATURDAY, SEPT 3

The headline news on Wednesday will be RALPH ROSE's gold medal and world-record-breaking throw in the shot put, and GEORGE POAGE winning bronze in the 400-yard hurdles—the first African-American to win an Olympic medal.

THURSDAY, SEPTEMBER 1 will be a day for connoisseurs of the eccentric, featuring the last Olympic outing for the 56-POUND THROW and the finals of the TUG OF WAR.

Friday will be a day off, but on SATURDAY, SEPTEMBER 3 there will be another nine ATHLETIC EVENTS to enjoy, as well as the weightlifting. The MEN'S TENNIS tournament, which began on August 29, will also reach its final, with the Boston blue blood BEALS C. WRIGHT triumphing in both singles and doubles events.

EXTENDED HIGHLIGHTS

For committed Olympian visitors we can offer a special EXTENDED STAY that will allow you to see many of the other sports contested.

The AQUATIC COMPETITIONS will be held in Exhibition Lake on SEPTEMBER 5–7. The east end will be reserved for life saving exhibitions from the US Coast Guard. Competitors will line up for races on a series of piers and rafts at the west end. The finishing line will be marked by a bamboo pole supported by a parallel float. The first day of competitions will see the late lamented PLUNGE FOR DISTANCE, in which competitors dive into the water and attempt to glide as far as possible without propulsion or surfacing. On the second day, try not to miss the controversial 50-YARD RACE, which will appear to be won by Hungary's ZOLTÁN VON HALMAY until the American J. SCOTT LEARY claims he had been interfered with. The judges, unable to agree on what has happened, will order a swim-off. Both men will commit foul starts before Halmay finally wins at the third attempt.

This will be as nothing compared to the fracas surrounding the DIVING COMPETITION on September 7. The Germans have brought their own diving board, wrapped in special coconut matting, and insist that scoring should be on the basis on acrobatic movements alone rather than the quality of entry into the water. The Americans disagree. At the end of the competition the German diver ALFRED BRAUNSCHWEIGER will refuse to compete in a bronze dive-off with the American GEORGE SHELDON, whom he feels he has already beaten. The German commissioner to the World's Fair, Dr. Theodor Lewald, who has donated a bronze statue for the winning diver, will be so incensed at the outcome that he will refuse to award it.

The ARCHERY TOURNAMENT (September 19–21) will be your only chance to see WOMEN (all six of them) compete at the Games.

Individual and team golds in GOLF (September 19–24) will be played for at the Glen Echo Country Club, which has recently completed building the first golf course west of the Mississippi. The all-American BOXING COMPETITION (September 21–22) sees gold medals fought over at seven weights. For those that can brave the cold outdoors, the WRESTLING (October 14–15) will be held in the middle of Francis Field Stadium at some distance from the stands; you'll need binoculars. Finally, for the diehards, the FOOTBALL competition will be concluded in late November with the experienced Canadian team GALT up against two local scratch sides, CHRISTIAN BROTHERS COLLEGE and ST ROSE PARISH. Galt will beat them 7–0 and 4–0 respectively to win gold. The play-off game on November 22 will go on and on at 0-0 as no provision for extra time or penalties has been arranged and it will eventually be called off for lack of light. The following day's replay will see Christian Brothers College win 2–0.

Sadly, there is no CLOSING CEREMONY. St Louis really could have done with Lionel Richie to party ("Karamu Fiesta Forever"), but that defining Olympian moment will have to wait until Los Angeles eighty years hence.

VE Day

MAY 7–8, 1945 ✵ LONDON

LONDON COULD TAKE IT. AFTER NEARLY six years of total war, rationing and living in shelters, having endured the Blitz and the V2 rocket bombs, it really was all over; and on May 8, 1945 the city came out to play and put on the biggest party it had ever seen. You'll arrive just in time to hear VE DAY declared and DRINK TO CHURCHILL'S HEALTH in the bohemian shadows of SOHO. On the day itself, you'll wake to the newspaper headlines at Piccadilly Circus, take VICTORY LUNCH AT THE SAVOY, and watch TRAFALGAR SQUARE fill to bursting before breaking out into drunken reverie and celebration. You'll hear CHURCHILL ADDRESS THE CROWD on Whitehall, watch KING GEORGE VI experiment with the royal wave from the balcony of Buckingham Palace amongst a crowd of 50,000, dance the conga all night, kiss a GI or two, and roll out the barrel.

> **NOTE:** WAG will shortly be offering a bespoke journey to New York for VJ Day on August 14, 1945. With the war in the Pacific over at last, this was the big one in the Big Apple. Join the crowds in Times Square and stand beneath the greatest wave of ticker tape the city has ever seen.

BRIEFING: PARTY LIKE IT'S 1945

It wasn't as if the country hadn't seen it coming. Germany's total defeat had been certain since the early months of 1945. The UK Board of Trade, ever cautious with frivolous consumption, had taken red, white, and blue bunting off the cotton ration. The Ministry of Works had announced—in advance of the end of the war and the anticipated celebrations—that "bonfires will be allowed, but the government trusts that only material with no salvage value will be used." While bunting and bonfires will be much in evidence on the night of May 7th, London still fretted as it waited for the final official signal that Germany had surrendered and that the War in Europe was over. Churchill had planned to announce this on the morning of May 7th but was forced to wait by the Soviets, who insisted that the surrender actually be signed and then the news embargoed before a simultaneous announcement was made. However, all of this was short-circuited by the release of the news of the surrender by the Associated Press in New York. With over a million people streaming onto the streets of Manhattan, it didn't make much sense to wait any longer and Churchill authorised a BBC broadcast, announcing that the next day, May 8th, would be VICTORY IN EUROPE DAY.

THE TRIP

You will be arriving at 6:00pm on the evening of MAY 7, 1945, in GLASSHOUSE STREET, a narrow passage that opens out onto the north side of Piccadilly Circus; please return here by 9am on the morning of May 9th for your DEPARTURE.

As you walk out into Piccadilly Circus it will be clear that London is still in wartime mode: the many neon and illuminated signs remain turned off and you will see the statue of Eros is boarded up and covered with adverts for war bonds. You may also sense the anticipation and excitement that are breaking out. Fighter planes have been executing outrageous loops and manoeuvres over the city centre all afternoon and the newspaper sellers have started shouting, "War is over!" Note, too, that toilet rolls have begun to fly out of office windows in central London; you may well see more.

V FOR VICTORY. THE SIGN CHURCHILL TAUGHT A NATION. IF YOU WANT TO JOIN IN, TAKE YOUR CUE FROM THE LADY IN THE TWEED JACKET.

MONDAY, MAY 7TH: EVE OF VE DAY

Take your time to explore the Piccadilly Circus area, but do try and be near a wireless in any one of the area's MILK BARS or CAFÉS for 7:40pm, when the OFFICIAL ANNOUNCEMENT OF VE DAY will be made. The BBC presenter will read the words: "in accordance with the arrangement made between the three great powers, tomorrow, Tuesday, will be treated as Victory in Europe Day and will be regarded as a holiday." You should then be able to hear the sound of thousands of TUGBOATS and SHIPS moored on the Thames blasting their horns.

As the news sinks across the country, PICCADILLY CIRCUS will be the place where the party begins. Over the next three hours around 10,000 people will gather here and in the nearby streets.

A BONFIRE will actually be assembled and lit on SHAFTESBURY AVENUE. A walk through Soho to the north will lead you to small fires set up in the ruined basements of bombed-out houses. This is a good chance, while there is still some breathing room, to join a CONGA LINE, do the HOKEY-POKEY, or dance with a range of servicemen. However, the party will be abruptly terminated by the weather. Around midnight a sharp summer THUNDERSTORM will break, lightning sheets will rent the sky, and the rain will fall for a couple of hours. Fortunately this is the last time you'll need to run for cover on this trip. The rest of your stay will be dry and May 8th will be a warm spring day. When you are ready for bed, head to the corner of Piccadilly Circus and Regent Street. A room has been booked in your name at the very respectable REGENT PALACE HOTEL (Piccadilly Circus, W1; Telephone—Regent 7000), reached across a courtyard graced by an Art Deco cut-glass dome.

EATING AND DRINKING

Britain has been on rations for nearly six years and they are not about to be lifted any time soon. Even in the West End of London, where restaurants, hotels, and cafés remain in operation, food can be hard to come by and is of the most variable quality. With this in mind we have taken the liberty of making a booking in your name for both lunch and dinner at the SAVOY HOTEL (90 The Strand, WC1; Telephone—Temple Bar 4343). It's a bit of a bargain. Despite its very well-heeled clientele, wartime restrictions mean both meals are a very reasonable five shillings (plus, for non-residents, a 3s 6d "house charge" sting). You will spot senior members of the allied military command and members of the Cabinet here and can enjoy the laid-back swing of American pianist Carroll Gibbons and the Savoy Hotel Orpheans. VICTORY LUNCH will be a three-course affair with wine: La Tasse de Consommé Niçoise de la Victoire, La Volaille des Iles Britanniques, La Citronette Joyeuse Deliverance, La Coupe Glacee des Allies, and Le Médaillon du Soldat. The VICTORY DINNER is less elaborate: soup, chicken, and iced peaches.

More-modest fare is available at the LYONS CORNER HOUSES dotted across the West End. These are palatial, too, in their own way and cater to everyone. Those of you who have chosen to

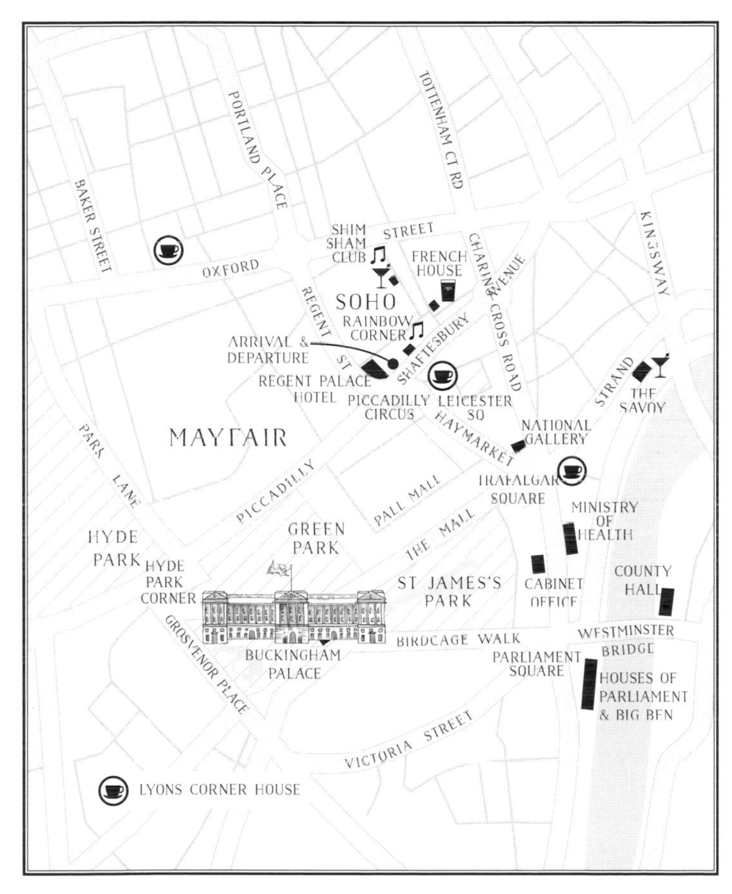

go in uniform may well be offered a free tea and a bun, but there is plenty to suit the most modest budgets. You will find a FOOD HALL on the ground floor and themed restaurants on the next three (or four in larger branches). You can use the phone, send a telegram, or get your hair and laundry done while you eat. The interiors are richly decorated with a kind of popular Art Nouveau swank that made them the people's palaces of the era. The waiting staff, all female, are dressed in identical black-and- white uniforms and are known as NIPPIES. Gay travelers will be pleased to find that an informal camp corner, orchestrated by the nippies, is available in these establishments. The classic corner houses are located on the Strand, Coventry Street, and Tottenham Court Road.

Hotel restaurants, with which the company has an arrangement, include the STRAND PALACE, the REGENT PALACE, and the CUMBERLAND. They are open twenty-four hours a day. For those with more contemporary tastes, some of London's first INDIAN RESTAURANTS can be found just a stone's throw from the main VE Day parties in Soho. Establishments include the DURBAR and the BENGAL INDIAN on Percy Street, the DILKUSH on Windmill Street, and the SHALIMAR on Wardour Street.

STREET PARTIES are another option for the peckish. Larders will have been raided, pantries emptied, and ingredients pooled by neighbors and families. There will be cakes and scones and things made of egg powder and grey pastry that only the brave will contemplate. Either way, go easy; people have been saving up for this for years.

More pressing may be the issue of ALCOHOL. Many establishments in central London will run out of beer relatively early on in the evening. We suggest you take the opportunity to drink up early and spend some of Friday evening exploring takeouts from the pubs and grocers in Soho.

TUESDAY, MAY 8TH: VE DAY

The morning of May 8th will begin brightly. Perhaps the first thing you will note is the sound of CHURCH BELLS ringing. They will be a ubiquitous background to most of the rest of the day. Piccadilly Circus is again a good place to begin and take in the moment. Newspaper sellers will be plentiful, passers-by will want to take a look at your copy and start up a conversation. The *Daily Mail* headline is "It's All Over," the *Daily Express* opts for "This is VE Day" and the US forces paper, *Stars and Stripes*, opts for the robust "Nazis Quit." The *Daily Mirror* has been teasing its readers with the prospect of cartoon starlet Jane appearing in the buff on the day the war was won. Pick up a copy and see how they handle that promise.

Alternatively, you might like to take the chance of a moment's peace and reflection before the party gets going. A SERVICE OF REMEMBRANCE will be held at ST PAUL'S CATHEDRAL at noon, but do try and get there in good time, as it's going to be completely packed. (Take the tube: eastbound Central line from Oxford Circus to St Paul's.)

VE DAY AFTERNOON

BUCKINGHAM PALACE—which is best approached from the Mall or Birdcage Walk—is not looking at its best. The stone is filthy and many of its windows are either blacked out or bricked up. The central balcony, which has been wreathed in crimson-and-gold velvet drapes, makes it look even worse. It's been a long war.

People will be gathering here from around 10am, their numbers growing steadily through the day, but there won't be much to see till gone 11am when various convoys of worthies, including the ROYAL FAMILY in OPEN HORSE-DRAWN CARRIAGES, make their way in and out of the Palace gates. The biggest crowds will be present for the appearances of CHURCHILL WITH THE ROYAL FAMILY at around 5:30pm and again at 9:30pm. The royals—KING GEORGE, QUEEN ELIZABETH, and the PRINCESSES ELIZABETH AND MARGARET—will make six other balcony appearances during the day. The king is in full naval officer gear, the queen has a terribly large hat on, which she will exchange for a diamond tiara for the late-night appearances. Elizabeth is the one wearing WVS khaki, Princess Margaret is in blue and exhibiting a nervous hair-touching tic. The princesses are going to be allowed out of the palace in the company of two Guards officers later this evening. If you do spot them, be cautious about following.

Just a few minutes' walk away along the Mall is TRAFALGAR SQUARE, which will be full by 1:00pm and remain that way into the small hours of May 9th. Of particular note are the HAWKERS working the crowd outside the National Gallery selling victory rosettes in red, white and blue, Union Jacks on sticks, cardboard party hats and Churchill badges. Listen out for the pitch, "Churchill for sixpence: Worth more!" In the afternoon a small ensemble will be performing selections from the best of GILBERT AND SULLIVAN OPERETTAS here. Look out later on tonight for the GIRL IN THE RED DRESS with white polka dots, who will be carried to the top of one of the fountains by two British army officers as the crowd applauds.

DANCING IN THE STREETS: WAR OFFICE GIRLS GET TO GRIPS WITH GIS.

If what you want are SPEECHES and moments of collective triumph, then you need to head for WHITEHALL, which leads off Trafalgar Square through Admiralty Arch, and PARLIAMENT SQUARE, which lies at the far end of Whitehall just beyond the Cenotaph. Along the way you will note the MINISTRY OF HEALTH, whose balcony will already be decked in Union Jacks, and which will be used throughout the day. Note also, two-thirds of the way down Whitehall on your right, the CABINET OFFICE and the rooms from where CHURCHILL will make his key 3pm broadcast. Finally, walk past the Cenotaph and you will see the HOUSES OF PARLIAMENT on your left. These locations will be jam-packed from about 1:30pm, and if you are to get a good vantage point we suggest you get down there well before

1pm. The crowd will be well ordered and there is no worry of crushes, but it is not for the claustrophobic. Do note the bus inching through the crowd, which has the words "Hitler missed the bus" chalked on it. A small cameo will occur at 2:40pm, when a naval officer will appear on the balcony opposite the Ministry of Works dressed as a PANTOMIME HITLER and proceed to impersonate him to much applause.

Then at 3:00pm Big Ben will strike the hour and the crowd will fall silent. You will finally hear CHURCHILL SPEAK through one of the many megaphones set up on Whitehall's lampposts: "Yesterday at 2:41am at headquarters, General Jodl, the representative of the German High Command, and Grand Admiral Donitz, the designated head of the German state, signed the unconditional surrender." The crowd will cheer especially on the announcement of the ceasefire, the relief of the Channel Islands, and the shout-outs for Eisenhower and the Russians. Churchill will wind up with a reminder about the Pacific War, still in progress, and end with: "Advance Britannia! Long live the cause of freedom. God save the king!" Then the OFFICIAL CEASEFIRE will be heralded by the bugles of the royal Horse Guards, followed by a mass rendition of the national anthem. Depending on where you are at this point, it might be possible to edge towards PARLIAMENT SQUARE, where at around 3:30pm you will see Churchill standing in the back of an open-top car which will be creeping through the ecstatic crowd, making the short journey from the Cabinet Rooms to the Houses of Parliament.

The BUSINESS OF THE HOUSE will take another hour or so with MPs gathering to hear Churchill, then attending a joint service of remembrance with the lords and a late departure for the prime minister after he forgets his cigars. CHURCHILL will reappear around 5pm puffing heartily on an absolute stonker as he is driven towards Buckingham Palace. Keep an eye on the skies (though you will hear the noise long beforehand) as a flight of Lancaster bombers will pass overhead to great acclaim, letting off enormous plumes of green and red smoke.

THE GANG'S ALL HERE—FAVOURITE SONGS

The sound track of 1945, especially on the dance floors, has been overwhelmingly American—with the big band and swing sounds of Glenn Miller, Benny Goodman, and the Andrews Sisters commanding the airwaves. However, in the crowd tonight it is the old sing-along favourites that have the floor. Expect to hear all of the these at some point on your trip. You might want to join in (song sheets can be found in your pocket or handbag).

Land of Hope and Glory
Roll Out the Barrel
Knees Up Mother Brown
We Going to Hang Up the Washing on the Siegfried Line
The Lambeth Walk
The White Cliffs of Dover
We'll Meet Again
Pack Up Your Troubles
It's a Long Way to Tipperary
Kiss Me Goodnight, Sergeant Major

ROLL OUT THE BARREL. A WAGON LOAD OF BEER ON ITS WAY THROUGH PICCADILLY CIRCUS (WITH EROS STILL UNDER WRAPS).

VE DAY NIGHT

One of the most exciting elements of the evening for Londoners is that the city will turn on its ILLUMINATIONS for the first time since the outbreak of the war. You may have noticed blackout curtains and papers being ripped from windows all day. This evening, street lighting will be in operation, public buildings of all kinds will be specially lit, and shops, theaters, and cinemas will turn on their neon signs. One particular moment of illuminated streetlife can be seen mid-evening on Haymarket, where the TIVOLI CINEMA's red signage bathes the street. You may see an ample woman in a Union Jack apron dancing with a middle-aged gent as an accordionist plays "South of the Border."

Though there is still much speech-making and balcony-appearing by the great and the good to come, the heart of the party is in PICCADILLY CIRCUS and the focus of the action is at RAINBOW CORNER. Previously the TROCADERO, a Lyons Corner House, this was taken over by the US armed forces in 1942 and has served as the centre of GI social life in London for three years. Indeed it has proved a magnet for locals entranced by the Americans' music, style, and demeanour. Today it will be party central as great waves of people move in and out of its bars and ballroom and back onto the street. From mid-afternoon, the club band will relocate to the balcony just above the entrance and provide an all-day sound track for dancers and revellers. GIs will be creating impromptu TICKER-TAPE PARADES from the windows in the upper floors, with toilet paper, files, phone books, apple cores, and shredded newsprint.

No traffic has been able to get through here since before 3pm, and from early evening the pace and the crush will be stepping up. Look out for the NORWEGIAN SAILOR stripped off at the top of a lamppost and a variety of British and American soldiers scaling EROS. Occasionally jeeps and taxis loaded with merrymakers, often piled onto the roof and the bonnets of the vehicles, will inch their way through. We do advise you not to join them. One

American soldier, already plastered in lipstick, will be calling out for volunteers to join his collection. Both sexes can expect the offer of a quick kiss and a cuddle to come from a variety of quarters.

If you want a break from the crowds in Piccadilly Circus, take a short walk to LEICESTER SQUARE. Its many trees will be beautifully lit and it is a good place to go for fireworks, flares, and a more raucous edge to proceedings. The streets of SOHO next door are even better. The FRENCH HOUSE on Dean Street is the place to get a glass of wine with London's FREE FRENCH, while Soho's émigré HUNGARIANS will also be partying. The WINDMILL THEATER on Great Windmill Street, whose risqué nude revue has been running since the 1930s, is always a place to go and see what's up and who is around. For those seeking something a bit jazzier and smokier, try the SHIM SHAM CLUB on Wardour Street.

BUCKINGHAM PALACE, too, will continue to be a focus for crowds during the evening, and as they get larger and larger ST JAMES'S PARK next door will become full to overflowing. For those who like a spot of celebrity-watching, keep in mind that NOËL COWARD will be putting in an appearance near the gates around 8:30pm. He will be accompanied by the entire cast of *Blithe Spirit*—his play running to packed audiences in the West End. The stars of the show include CECIL PARKER, FAY COMPTON, and MARGARET RUTHERFORD. The noted popular composer IVOR NOVELLO will also be with them.

At 9pm KING GEORGE VI will be broadcasting his address to the nation live from the Palace and will then make a final BALCONY APPEARANCE WITH CHURCHILL and his family. You will find that by this hour things have got altogether more raucous in front of the Palace gates. Look out for the young trumpeter HUMPHREY LYTTELTON, who will be exuberantly playing on the Victoria Monument in front of the main entrance to the Palace. He will be joined by a man with a large bass drum strapped to his stomach, an American sailor with a trombone and an old lag blowing a single but effective bass note from the horn of a gramophone. After twenty minutes dancing and playing, Lyttelton will strike up "High

"WERE WE DOWNHEARTED?" CHURCHILL SOAKS UP THE ADULATION, ALONGSIDE THE DUO—BEVIN (RIGHT) AND ATTLEE (FAR RIGHT)—WHO WILL SOON REPLACE HIM.

Society" and be lifted onto a handcart. The whole ensemble will then make its way over the next couple of hours down St James's Street to Piccadilly Circus, then to Trafalgar Square and back again to the Palace. ST JAMES'S PARK is also worth a wander at this hour. The trees and paths will all have been lit and one of the city's largest bonfires is roaring, while in the park's darker corners you will find the atmosphere amongst couples unrestrained to say the least.

The final balcony appearance of the night will be CHURCHILL AT THE MINISTRY OF HEALTH on Whitehall. He will be preceded by members of the cabinet, including Deputy Prime Minister and leader of the Labour Party CLEMENT ATTLEE, and the president of the Board of Trade, ERNEST BEVIN. Listen out for the drunken Tories and aristocrats who will be giving them some stick and calling out for Churchill. When he does arrive, Churchill will conduct more of a call-and-response dialogue with the crowd rather than give a speech: "We were all alone for a whole year. There we stood. Did anyone want to give in?" The crowd will roar back, "No!" "Were we downhearted?" will be met by the call, "No fear!" The whole thing will wrap up about 10:30pm with the Prime Minister leading the crowd in singing "Land of Hope and Glory."

AWAY FROM THE WEST END

Although the West End is the heart of the action, there are many smaller and more intimate celebrations going on in the suburbs, which might be worth your time.

NORWOOD The bonfires and effigies at street parties in this south London suburb are particularly good. Apsley Road boasts a Hitler swinging from a gallows so large that it effectively blocks the road. Nearby Balfour Road has Hitler dressed up with a swastika flag and a sign reading "I have no further territorial claims in Europe." "No" has been crossed out and "Hell" painted over "Europe." They'll be BURNING THE FÜHRER around 8pm.

London and Croydon Railways train to South Norwood.

RAVENSCOURT PARK There will be a huge bonfire in this small park, between Hammersmith and Chiswick. A sizable band will also be playing throughout the evening and you can enjoy the bizarre sight of the massed ranks of uniformed nurses from the nearby Queen Charlotte's Hospital dancing on the lawns.

Piccadilly line to Hammersmith then District line to Ravenscourt Park.

ST PAUL'S Although it is not the centre of festivities, the great dome of Wren's Cathedral is a focus of the city's best illuminations. The crossed searchlights over the golden cross on top of the dome are particularly striking. Later in the evening a spectacular "V for Victory" illumination will be created over the dome.

Central line to St Paul's or Bank.

WILLESDEN The teenagers of Hanover Road in Willesden have been particularly active building their bonfire. You will see handwritten bills on the lampposts of the surrounding roads as you walk towards the party. A bombed-out plot, strung with Union Jacks and bunting, has been piled high with salvaged wood, while a remarkably life-like Hitler swings above the pyre on a set of gallows. A stuffed model of Goering with two iron crosses on his uniform sits on a chair at its foot. At 9:15pm a crowd of 100 teens will have gathered and a gramophone will have been brought outside. At 9:30pm the bonfire will be lit, to cries of "Don't let him end too soon, let him linger"; some of the kids will douse Hitler with a hose to prolong his agony. Expect a lot of old fireworks and home-made flares. Then a piano will appear in the street and you will get all the classics. The dancing will be shy to begin with, but by 11pm the place will be roaring.

Bakerloo line to Queens Park.

MIDNIGHT AT BIG BEN

As the evening inches towards midnight and the official end of the War in Europe, you might like to think about getting to WESTMINSTER, where BIG BEN will be striking the hour and an enormous Union Jack flying above the House of Lords will be dramatically picked out by multiple spotlights.

The best vantage point is probably in the middle of Westminster Bridge, which lies on the south side of Parliament Square. From here you will be able to see not only Big Ben and Parliament, with its river terrace garlanded by lights, but the south bank of the river. Here you will see rotating searching lights, the seat of the London County Council, COUNTY HALL, picked out in red, white, and blue lights, while the boats and ships on the Thames will all be strung with colored bulbs.

As midnight approaches, silence will fall over the considerable crowd, broken only by the final bong of the midnight chimes. Expect cheering, fireworks going off all over the place, searchlights whirling, and every boat on the Thames blasting its horn. If you do find yourself in a quieter spot at midnight, the radio will certainly be on. Listen out for the BBC's Stuart Hibberd announcing: "As these words are being spoken, the official end of the war in Europe is taking place."

At this point you may wish to just carry on and party through the night, rather than making use of your hotel. You will have no shortage of fellow revellers. For those who need a brief lie-down, GREEN PARK, HYDE PARK, and ST JAMES'S PARK will also be available. The evening will stay dry, the temperature will be cool but not cold. There will plenty of fires to keep you warm and it will be safe. Police reports confirm almost zero crime in the city today.

Woodstock Festival

AUGUST 15–18, 1969

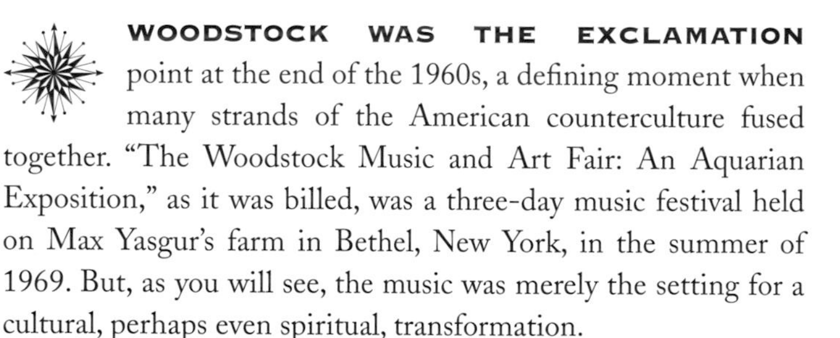

WOODSTOCK WAS THE EXCLAMATION point at the end of the 1960s, a defining moment when many strands of the American counterculture fused together. "The Woodstock Music and Art Fair: An Aquarian Exposition," as it was billed, was a three-day music festival held on Max Yasgur's farm in Bethel, New York, in the summer of 1969. But, as you will see, the music was merely the setting for a cultural, perhaps even spiritual, transformation.

Planned with 100,000 people in mind, half a million showed up and walked across the woefully incomplete fencing that the organisers had failed to erect. At 4pm on Friday, as they surveyed the great mass of humanity that had gathered before the main stage, the management declared it a free festival. Not long after, New York governor Nelson Rockefeller designated it a disaster zone.

The latter is pretty descriptive. The event is absurdly undercatered and underprovisioned, a state compounded by the roads being blocked for miles around and by squalls and storms, turning much of the site into a MUDBATH. The musicians arrive late, the schedules are chaotic, and drugs of all kinds are plentiful. But something even more powerful than the orange sunshine acid doing the rounds is at work: a massive outbreak of spontaneous sharing, giving, and caring breaks out.

And through it all the most fabulous array of the era's musicians perform: from the Latino rock of SANTANA to the multicultural funk of SLY AND THE FAMILY STONE, from the protest songs of JOAN BAEZ and JOHN SEBASTIAN to the laments and sweet harmonies of CROSBY, STILLS, NASH AND YOUNG. And, of course, the showstopping finale from JIMI HENDRIX. The organisers had promised three days of music and love, and though things didn't pan out quite as they had intended, that is what the "half-a-million strong" Woodstock Nation got.

BRIEFING: WHY WOODSTOCK?

Woodstock was the brainchild of four twentysomethings: the enigmatic MICHAEL LANG, an über-cool head-shop owner and small-time music promoter; the maniacal ARTIE KORNFELD, who was vice president of Capitol Records; and two very straight Manhattan flatmates playing venture capitalists, JOEL ROSENMAN and JOHN ROBERTS.

The quartet's initial encounters turned on the idea of establishing a hip new recording studio in the town of Woodstock in upstate New York, which many of the era's leading musicians—like Bob Dylan and The Band—had made their base. The never-built studio was to be funded and promoted by a big music festival. Building on the success of San Francisco's huge HUMAN BE-INS and the MONTEREY POP FESTIVAL, both held in 1967, the project soon mutated into something much bigger and acquired a roster of musicians that was truly extraordinary.

PLAN A had been to hold the event in an old industrial park in the town of Wallkill, about thirty-five miles from Woodstock. Just weeks before the opening day, however, with many tickets already sold, local protests forced the land owner to withdraw his offer. A series of chance encounters led the organisers to the hamlet of WHITE LAKE near Bethel, New York, and an unlikely saviour in the form of dairy farmer MAX YASGUR, renowned for his sour cream, yogurt, and chocolate milk. A deal was struck, and in a mad flurry of activity in late July and early August 1969, Yasgur's farm was converted into a huge festival site.

It was staffed by the HOG FARM, a hippie commune from New Mexico who would run the kitchens and freak-out tents and, together with a bunch of off-duty New York police officers, provide the PLEASE FORCE, Woodstock's very own security force trained to say, "Please don't do this, do something else."

By opening day, Friday, August 15, 1969, the stage was built, the sound system was just about working, and 50,000 people had already turned up, most of them without tickets. They just pitched their tents and sat waiting for the music to begin. Meanwhile, the festival audience continued to grow.

THE DUDES WHO MAKE WOODSTOCK: MICHAEL LANG AND ARTIE KORNFELD. THEY LOSE THEIR SHIRTS BUT IT ALL TURNS OUT COOL.

THE TRIP

Your point of arrival and departure is the inside of a green and white 1963 VOLKSWAGEN CAMPER VAN, parked on the north verge of Route 17 about half a mile from BETHEL. As you will see as you get out of the van, it is just one of thousands of vehicles parked up on this narrow stretch of road that curves through the low hills and woods of the Catskills. The centre of the road will be filled by a steady stream of festival-goers abandoning their cars and walking west. Take a breath and enjoy the sweet smells of fresh cow manure and Colombian Gold that fill the air. The first major turning on your right takes you up into Bethel proper and to the hamlets that cluster around WHITE LAKE. You will see that cars have begun to pile up here too, and the traffic is blocked to the north. The locals' attitudes towards "the kids" are very variable. Some have created impromptu stores in their front gardens; others will offer you a cup of coffee and something to eat, particularly once the event has been declared a disaster zone and is all over national TV.

For those who like to plan ahead, take the short walk up Route 14 to Kauneonga Lake, then left onto Broadway, where you will find a considerable crowd around VASSMER'S GENERAL STORE. Featured in the movie *Woodstock*, this shop is run by the genial Arthur and Marian Vassmer. While many local stores have closed, Vassmer's will serve thousands over the long weekend. This is your best chance to stock up on chocolate, cookies, wine, cigarettes, Rizlas, and toilet paper—the standard fare of today's festivals, but lamentably absent from Woodstock.

THE FESTIVAL SITE

"By the time we got to Woodstock, we were half-a-million strong," Joni Mitchell will sing in her famous paean to the festival. Now it's time to join them. Keep heading west on Route 17 and go

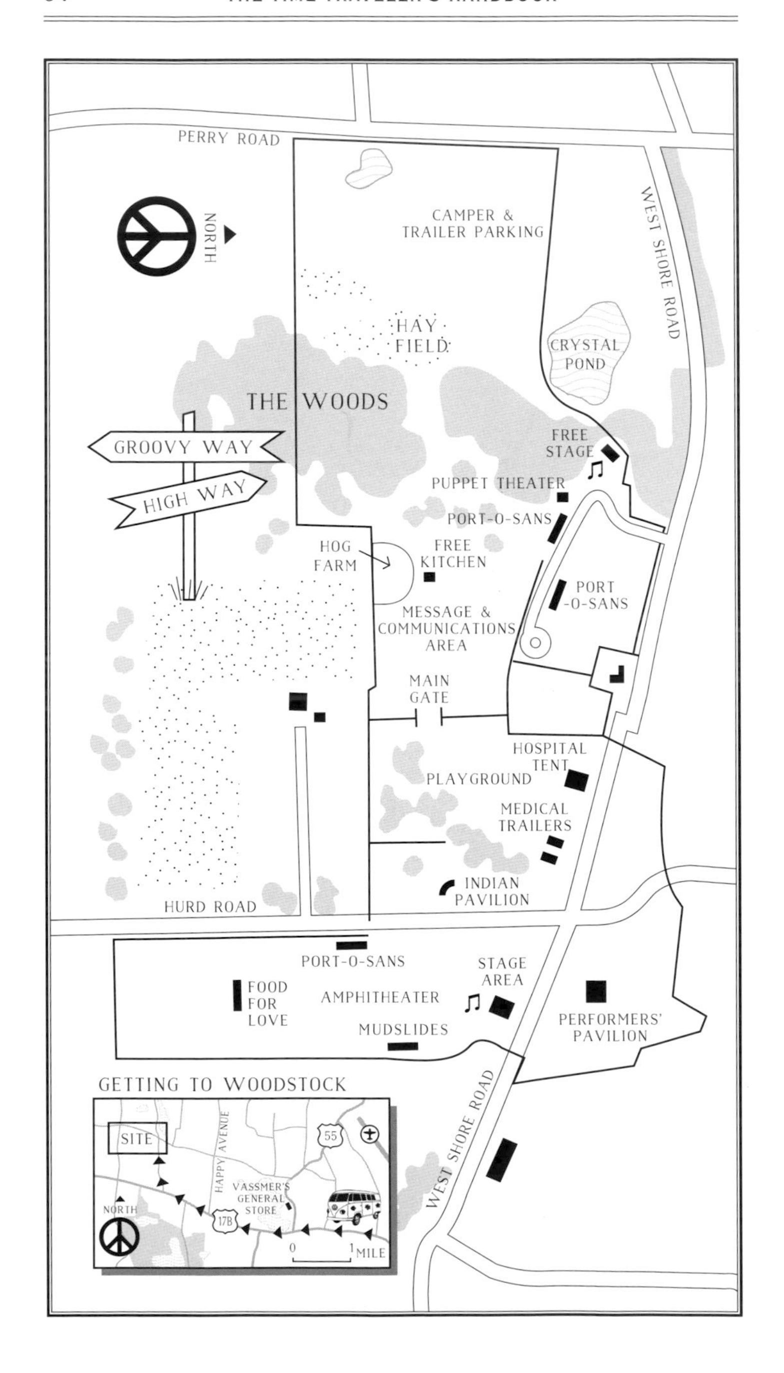
PERRY ROAD
NORTH
CAMPER &
TRAILER PARKING
WEST SHORE ROAD
HAY
FIELD
CRYSTAL
POND
THE WOODS
GROOVY WAY
HIGH WAY
FREE
STAGE
PUPPET THEATER
PORT-O-SANS
HOG
FARM
FREE
KITCHEN
PORT
-O-SANS
MESSAGE &
COMMUNICATIONS
AREA
MAIN
GATE
HOSPITAL
TENT
PLAYGROUND
MEDICAL
TRAILERS
INDIAN
PAVILION
HURD ROAD
PORT-O-SANS
STAGE
AREA
FOOD
FOR
LOVE
AMPHITHEATER
MUDSLIDES
PERFORMERS'
PAVILION
WEST SHORE ROAD
GETTING TO WOODSTOCK
SITE
HAPPY AVENUE
55
VASSMER'S
GENERAL
STORE
NORTH
17B
0
1 MILE

round the south end of White Lake. There won't be any doubt as to where everyone's going. Your only decision is when to cut off to your right towards the festival. We recommend that you head down to the valley about two miles out of town, just before the road crosses a stream. Then, keeping the wooded area to your left, it's a ten-minute stroll across the hayfields. Try to remember where you've turned off. Fix it in your mind. Really. You have to get back to this road and then to the VW camper on Monday morning and you may not be in peak condition when the time comes.

THE CAMPSITE

You should now be looking down on the main designated CAMPSITE. This lies outside the main festival arena and what remains of the fence surrounding it, but with 50,000 people estimated to have arrived by Thursday night, the campsite will already be spilling over into the woods. Grab what space you can and look to see if there are still any piles of cut logs around. The Hog Farm left them out for campers to take and the organisers encouraged people to make fires and get cosy. Feel free to do so.

TOILETS AND WASHING FACILITIES are thin on the ground and not for the faint of heart. If you are wise, you will have picked up a stash of tissues at Vassmer's. In addition to the Port-O-San at the bottom of the hill, there are also toilet blocks behind the Indian Pavilion in the festival arena and at the top of the hill in front of the main stage, next to the food concessions. For the hygienically minded the best bet is the large and relatively clean FILIPPINI POND, a fifteen-minute amble due north of the Free Stage. Skinny-dipping, diving, and frolicking are highly recommended, though not compulsory.

Depending on how long it has taken you to get this far, you may find that an area in front of the main gate has started to become the INFORMATION AND COMMUNICATION ZONE—an improvised necessity in a world without the mobile phone. It is a cluster of notice boards, messages on totem poles, and tents with helpful people in them. You may not find who you are looking for but you are bound to run into someone or something interesting here all weekend.

ABBIE HOFFMAN, the notorious and theatrical political activist, has persuaded the organisers to let him have his own tent in the Communication Zone. From here, he and other Yippie (Youth International Party) activists will be issuing hastily typed INFORMATION SHEETS to the bamboozled crowd. The Saturday-morning edition, for example, will read as follows: "Everything might seem groovy now. But think about tomorrow. Life could get hard. If you're hip to the facts below, pull together in the spirit of the Catskill mountain guerrilla and share everything. Dig It!" There really isn't enough food at Woodstock, so take your cue from the man: share what you have and receive gifts graciously. Water is not in short supply, with standpipes widely available, though the functioning of the system is at best erratic. Contrary to the rumours you will be hearing, neither the tap water nor the bottled stuff flown in by the National Guard is laced with LSD.

WEATHER REPORT

Your walk to the festival site will be in very pleasant late summer sunshine with just a hint of cloud. But don't kid yourself: this is Woodstock. It's going to rain and you've got to love it. Richie Havens, who opens the show on Friday, will later say, "The rain made the people interact with each other ... to share whatever we had ... so I balance it out as a cosmic accident." Many of your fellow festival-goers will be in JEANS AND T-SHIRTS or less. While this is perfectly permissible, the agency does recommend some kind of period-appropriate RAINWEAR (or make use of the raincoats and micro-tent stashed in the camper van, next to the cooker).

On FRIDAY the delightful mix of sun and scurrying clouds will turn blustery later in the evening. At around 10pm, during Ravi Shankar's set, it will start raining heavily. Lighter showers will persist through the night, turning to steady drizzle early on SATURDAY morning. Things will dry for out for the first few acts but another rain storm after lunch will continue until after 4pm. When the music resumes, temperatures will be a warm but damp 70°F. Rain returns for an hour or so later in the evening, peaking during the

Grateful Dead's set. This will be cut short for various reasons, including the risk of the band getting electrocuted as they play in puddles of water.

SUNDAY will begin sunny, warm, and breezy, but storm clouds will start to gather at around 2pm. Joe Cocker will have just completed his set at around 4:30pm, when the final thunderstorm will break, drenching the festival for an hour. The rest of the day will be dry and, for those that have the stamina to stay, MONDAY MORNING, post-Hendrix, will be sunny and golden.

"WHAT WE HAVE IN MIND IS BREAKFAST IN BED FOR 400,000. KEEP FEEDING EACH OTHER ... AND IF YOU'RE TOO TIRED TO CHEW, PASS IT ON."

EATING AND DRINKING

The food concession arrangements at Woodstock are unsatisfactory, to say the least. Unable to secure the services of established caterers, the organisers have gone with an outfit called FOOD FOR LOVE, long on promises but short on capital. Despite having been loaned $65,000 and had their stands built for them at the top of the hill overlooking the the main stage, Food for Love will run out of supplies—a basic range of burgers, hot dogs, and sodas—within a few hours.

If you do want to get a burger, get there on Friday late afternoon, after Woodstock has been declared a free festival, when a group of leftist activists from Greenwich Village will put in an appearance, chanting, "Free food for the people" and looking threatening. A squad from the Hog Farm/Please Force will appear (note their motley red armbands), giving out large burning incense sticks to calm everybody down. You will then see that most of the people at the concession stands are exchanging joints for burgers and the entire kitchen crew are smoking them at a lick. For the next hour or so they will just hand out food for free. But get in line pronto, for soon it will all be gone.

Your most reliable food source for the weekend is the FREE KITCHEN, located just to the west of the Information and Communication Zone. They will serve up three meals a day all through the festival. Breakfast is a muesli mix of rolled oats, sesame seeds, honey, raisins, and wheat germ. Lunch and dinner consist of great vats of bulgur wheat and brown rice served up with whatever locally sourced vegetables can be found. There are some mighty soy sauce dispensers, too. The queues are long but brisk.

Woodstock's greatest eating pleasures, however, are the RANDOM OFFERINGS that will come your way. From Saturday onwards the helicopters of the National Guard and medical services will be introducing a strange array of canned goods into the mix, including thousands of one-ounce cans of olives and tins of tuna. Good luck with the can opener. Saturday afternoon will see packages of baloney sandwiches, Hershey bars, Melba toast, and plastic bottles of cola thrown down to the crowd. Locals in Bethel and Monticello will also take it upon themselves to bus in loads of home-made sandwiches and trays of hard-boiled eggs.

THE ARTS FESTIVAL

Everyone will remember Woodstock for the music, but it was always intended to be a wider arts festival. Not much of it has survived the late changes of plan, but the best of what remains can be found to the east of the main campsite in and beyond the woods.

Nestled in the trees near the West Shore Road is the FREE STAGE, built and vaguely curated by the MERRY PRANKSTERS, a troupe of California acid-head situationists given to impromptu happenings and drug-fuelled partying. The Pranksters' fabulous painted school buses can be seen parked up behind the Free Stage. Performances include yoga sessions, a late-night appearance on

Friday by Joan Baez, and spiritual and meditative performances with a massed array of Tibetan gongs. This is the best part of the festival for joining improvised groups of drummers—still very much a novelty. You also stand a good chance of being offered hallucinogens.

Additional entertainment is available about 200 yards up the slope, where the PUPPET THEATER is putting on shows all day. Push on through the trees from the Free Stage and you will come to what was intended to be the car park. You will also find it functioning as an overspill campsite and general hangout, especially the HAY FIELD up the hill and the shady areas around CRYSTAL POND. However, the most interesting part of the site is THE WOODS. Covering about four acres, this delightful strip of woodland is criss-crossed by two main paths, signed as THE HIGH WAY and THE GROOVY WAY, and lit by long strings of lights. Along the paths you will find an eclectic assortment of tarot readers, troubadours, meditators, frolicking couples, bead sellers, and dope dealers.

Most of the crowd will be heading east towards the main stage. As you follow through the main gate or over the now-trodden-down fencing, take note of the cluster of tents and the large white trailer parked at the bottom of the hill. The HOSPITAL TENT is the main place to seek medical care. You may have noticed that many of your fellow festival-goers are barefoot and quite a few of them are pitching up here with glass and metal cuts. There will also be thousands of bad-trip casualties to be attended to over the course of the weekend. A similar "talking down" service is also available at Hog Farm from late on Friday evening (see TURNING ON, TUNING IN, AND FREAKING OUT, page 68).

Two special attractions in this area are the INDIAN PAVILION and the PLAYGROUND. The former is a collection of tents and tepees exhibiting the best in contemporary Native American art, with many artists from New Mexico and California in residence. The Playground consists of a selection of wood-and-rope structures dotted around the field and designed to be played on. Look out for the large stone-on-a-rope swing and the maze. Best of all, ascend the tree-trunk climbing frame and throw yourself onto the fabulous mountains of hay bales below.

TURNING ON, TUNING IN, AND FREAKING OUT

Given the sharing ethos of the festival, you should have no problem getting yourself just as stoned as you want to be. Expect joints and pipes to be passed your way and feel free to make polite inquiries. Possession of a cigarette lighter, useful during key moments of the music, is invaluable as your way to the REEFER: most will have forgotten or lost theirs. For those of you who would like to secure your own supply, we believe the going rate is around $5 for a bag of something nice from Colombia.

There is no shortage of LSD at Woodstock either, most of it in pill form. The Pranksters, who are running the Free Stage in the main campsite, will be doling it out for free through most of the weekend. The famous announcement from the main stage warning about the brown acid being "not specifically too good" has some truth, but as the man says, "It's your trip." Reports of green, blue, and red acid abound. Casual inquiries of those not too far gone will get you to them. Usual rules apply: just take half a tab and see how it goes; you can always go back for seconds.

A range of strong prescription drugs—amphetamines, Valium, and other benzodiazepines—magic mushrooms, and psychoactive cacti will be in circulation. The MESCALINE going round appears to be particularly rough—if swirling Aztec imagery and icons is not your thing, it is perhaps best avoided. ALCOHOL is available, and wine bottles will be coming your way pretty often, but if this is your drug of choice we suggest you make sure you visit the liquor stores in Bethel early on during your stay.

On Friday afternoon you will find MEDICAL FACILITIES at Woodstock rather thin on the ground. At the north end of the main festival area there are a cluster of first-aid tents, and the small medical trailer parked alongside them houses trained medical staff. As the number of psychedelic freak-outs rises sharply over the weekend, CHILL-OUT TENTS will be established alongside the original hospital tents and then around the festival. Many of these will have characteristic pink and yellow stripes, which may or may not be soothing to the target market. Others will be obvious from their clientele. The regular medical staff on site will be reinforced by new arrivals from New York helicoptered in on Saturday morning. If you go to them on an acid bummer, they will give you a big shot of Thorazine. This will stop the trip, but send you into something close to a vegetative state.

For those who prefer talking cures, there is an alternative. Most of the chilling out and calming down work will be done by members of the Hog Farm, supplemented by people they have brought down from their own bad trips who have become impromptu psychedelic paramedics. Here the advice is to ride the thing out and enjoy the beauty. Saturday evening is a particularly good time to have your freak-out, as John Sebastian (who will have played a set on the main stage by this point) will come with the members of the Lovin' Spoonful to play a soothing acoustic set to the assembled acid casualties.

THE UNDRESS CODE. THIS IS PERFECTLY ACCEPTABLE, BUT THE COMPANY DOES RECOMMEND SOME KIND OF FOOTWEAR.

DAY BY DAY: THE MUSIC

The main stage is situated in a naturally shaped amphitheater, surrounded by a huge crowd. To see this at its best, approach over the brow of the hill by the service road. At the foot of the hill you will see the MAIN STAGE, thirty yards wide but tiny from this distance. Around it is a high wooden garden fence, and behind it is the performers' backstage zone. It's worth dwelling for a moment on this city of tents and tepees and the fantastic curving wooden bridge that carries the musicians and crew to and from the stage. Note also the sixteen loudspeaker arrays on 70-foot-high towers that snake up the hill. You will see that many people have climbed them for a better view of the stage. There will be no reports of casualties, but we do not recommend joining them.

This is the moment to take in the sheer scale of the crowd. There will be anything up to 300,000 people in front of you. Of course, the demographic is mainly white and middle class, with a heavy bias towards New York, but watch the passing throng and you will see all manner of young folk: African, Native and Asian Americans, college students and dropouts, hippies, Yippies and folkies, politicos and preppies, and Hare Krishna monks.

This is also the time to think about where you want to sit to watch the music: close up to the fence, mid-distance, or way at the back? Wherever you choose, the sound quality will be remarkably good, given the conditions. And, to hear the bands, you will need to stick with your choice for the day. While the music is on and the sun is out, there won't be room to swing a cat. If you need to, now is the time to head for the concession stands at the top of the hill and the nearby Port-O-Sans. This area is also where you will find some of the best MUDSLIDES and puddles once the rain arrives.

FRIDAY, AUGUST 15TH

On Friday, the crowd will have been gathering in front of the main stage since morning, waiting for the concert to begin just after lunch. You may well feel a certain restiveness as the hours slip by without any musicians appearing, but rest assured the show will go on.

At just gone five o'clock, CHIP MONCK, your MC for most of the next three days, will make the announcement everyone has been waiting for: "Sit down! Stand up! Do whatever you wish to do but we're ready to start now and I bet you're pleased with that. Ladies and gentlemen, please, Mister Richie Havens." RICHIE HAVENS and those members of his band who have been able to get to the gig are being literally pushed onto the stage as the only outfit even close to being ready. Note the bass player arriving just after Havens; he has just finished an epic twenty-mile walk through the traffic jams carrying his guitar. Havens will hold himself and the crowd together with a two-hour performance of improvised hard-edged folk and protest tunes, including an incomparable rendition of "Freedom."

Behind the scenes, the producers are desperately searching for acts that are both physically and mentally present. Serendipitously, they have found the Indian yogi and guru SWAMI SATCHIDANANDA hanging out backstage. He is up next, sent on to play for time. Around seven o'clock you should hear his steady high-pitched voice even if you can't see this tiny figure in saffron robes with exuberant long ringlets and a magnificent white beard and moustache. Sitting cross-legged on a small podium and surrounded by his monkish devotees, Satchidananda will declare, "Through music we can work wonders. Music is the celestial sound, and it is sound that controls the whole universe, not atomic vibrations. Sound energy, sound power is much, much greater than any other power in this world." Right on, brother! His set will conclude with a collective chant of "Hari-Om."

Just after half past seven the show proper will finally be ready to roll. SWEETWATER, originally booked to open the day, are first up,

LEVITATION IS ALL YOU NEED. SWAMI SATCHIDANANDA OPENS PROCEEDINGS ON FRIDAY AFTERNOON.

mixing folk rock with cello and flute. At 8:20pm the frizzy-haired BERT SOMMER is the first of a pair of solo folk-guitar boys, his current hit "Jennifer" and a brilliant version of Simon and Garfunkel's "America" evoking the biggest responses. He is followed an hour later by the willowy TIM HARDIN, a rising star of the folk circuit. According to witnesses backstage, Hardin has been "completely blitzed" for the past twenty-four hours and is in profound dread of the enormous crowd. But, despite his shakiness, he will manage to take things up a notch with his version of "Simple Song of Freedom" and his current hit, "If I Were a Carpenter."

Things will be taking a sharp turn, musically and climatically, around ten o'clock. RAVI SHANKAR, the great Indian sitar player, will take the stage and rip through three standards from the Indian classical repertoire before the heavens open and the first of many short, sharp downpours begins. Once the shower stops at around 11 o'clock it will be time for the recently launched New York folkie MELANIE to play. Just prior to her set, the stage announcer will say, "This is the largest crowd of people ever assembled for a concert in history, but it's so dark out there we can't see and you can't see each other. So when I say 'three' I want every one of you to light a match." Everybody who can get it together will do so, though you stand less chance of getting your fingers burned if you've taken our advice and brought along a lighter. Melanie will be sufficiently inspired by the illuminations to write the hit song "Lay Down (Candles in the Rain)."

At five minutes to midnight a very young and very stoned ARLO GUTHRIE will take to the stage. The son of American folk hero and protest-song pioneer Woody Guthrie, Arlo has spent the past five hours hanging out in the crowd, having a smoke, and expecting to play tomorrow. The fact that he is on now has thrown him somewhat. There will be some sticky moments and lapses of memory during his set, but the stoner paranoia of "Coming into Los Angeles" and the optimistic resilience of "Amazing Grace" should make a lot of sense. Finally, at just after one o'clock in the morning, the last act of the night will be tuning up: the queen of the American folk scene, JOAN BAEZ, six months pregnant at the time. With the weather alternating between drizzle and showers, Baez's rendition of "We Shall Overcome" may feel particularly pertinent.

SATURDAY, AUGUST 16TH

First up on the main stage is Tom Law from the Hog Farm, enjoining you and everyone else to take part in a vast meditation and yoga session. Muesli, served from huge plastic vats, will be available at the left-hand side of the stage from about 11am. Saturday's music will begin quietly with the little-known QUILL playing a short set from about 12:30pm, enhanced by the many maracas and other percussive instruments that they will be handing out to the crowd. Feel free to grab yourself one, but do remember not to bring it home lest it should interfere with the time–space continuum.

The failure of a number of artists to have shown up even by this advanced stage will ensure a special treat for early risers. At 1pm COUNTRY JOE MCDONALD will be performing solo. The high point will come at the end of the set when McDonald will orchestrate the Fish Cheer: "Give me an F! Give me a U! Give me a C! Give me a K! What have you got?" Feel free to join in. This will segue into the fabulously upbeat "I-Feel-Like-I'm-Fixin'-to Die-Rag," a song that combines the sardonic black humour of Tom Lehrer with the politics of the anti-war movement. This will be the first major collective act of singing.

You may wish, like many in the audience, to time your recreational activities so you peak at around 2pm today, when SANTANA will be playing. One of the musical highlights of the weekend, the band's blend of raging rock guitar, keyboards, and percussive Latino rhythms makes for an ecstatic performance, with a matching response from the crowd. "Soul Sacrifice" is perhaps the first amongst equals. Of special interest, dancing close to the stage, is the naked man with the sheep in his arms. He has been walking around the festival with the said animal in his arms for most of the past thirty-six hours.

The light rain will cool you down after Santana, but the sun will return as JOHN B. SEBASTIAN takes to the stage in his wire-rim glasses and paisley shirt at around half past three. Previously the lead singer of the Lovin' Spoonful and currently beginning an ill-fated solo career, John B's set will capture the moment perfectly. As Woodstock producer John Morris will later recall, "Something magical transformed the stage when John Sebastian ambled out." His short, gentle acoustic set will lift you and the crowd, your clothes will be drying out, and Sebastian, as stoned as his audience, will forget the lyrics to a number of his tunes but find the words that are needed. When he looks out, in real awe, at the scale of the crowd and says, "Wow—We really are a whole *city*," the crowd will rise as one. Then it will be onto the

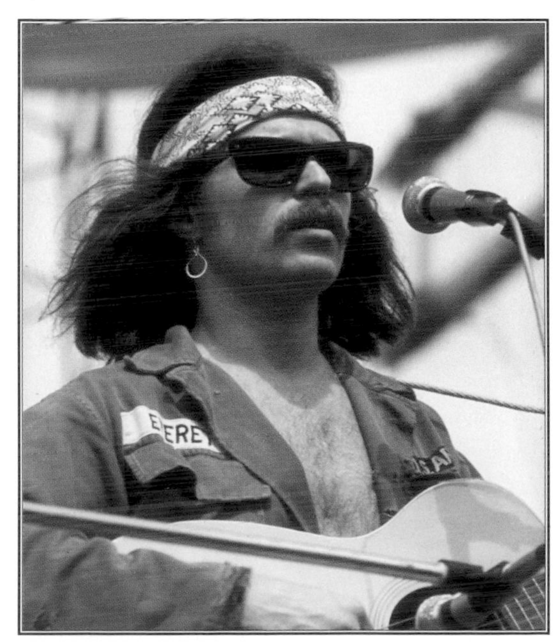

"HOW DO YOU EXPECT TO STOP THE WAR IF YOU CAN'T SING LOUDER THAN THAT?" COUNTRY JOE HAS A POINT. FEEL FREE TO JOIN IN.

blues rock of the KEEF HARTLEY BAND, followed by the Indian-folk-jazz fusion of the INCREDIBLE STRING BAND. Most of the crowd seem confused by the cult British hippies' modal improvisation and unusual tunings.

More mainstream service will be resumed at half past seven, with a solid set of foot-tapping boogie provided by CANNED HEAT, enlivened by the high-falsetto blues of lead singer Alan Wilson on "Going Up the Country" and "On the Road Again." Midway through the set a rogue member of the audience will clamber up on the stage and embrace him, before the pair share a cigarette and dance together. MOUNTAIN, on at nine, will be offering straight-ahead rock. You, like the band, will be beginning to battle the drizzle that is turning into rain, but hang in there: at half past ten the GRATEFUL DEAD are coming on. The band have spent the past two hours embroiled in a gigantic row over getting paid in cash up front, during which time the stage has morphed into one vast puddle. Certainly when recollecting the night, the group will not claim it to have been their finest hour, but for many the oddball noodling, jamming, and rifling amongst short-circuiting electrical equipment seems a pretty standard night out with the Dead. Things will come to an abrupt halt during their fifth song, "Turn on Your Love Light," when the amps will overload and switch off.

The rain will calm down for the rest of the night, allowing an amazing succession of artists to take the stage in the early hours of Sunday morning. Do not miss CREEDENCE CLEARWATER REVIVAL performing a storming set of their Louisiana swamp blues—it won't feature in the movie of the festival, as the band refuse to be filmed. JANIS JOPLIN AND THE KOZMIC BLUES BAND will be clearly the worse for wear, having consumed an enormous quantity of champagne, but Janis's slurred blues howl on the encore "Ball and Chain" will be electrifying. Half past three brings SLY AND THE FAMILY STONE, California's freewheeling multi-ethnic alchemists of soul, psychedelic rock, and funk. You will notice by now that the vast majority of the crowd have curled up, flaked out, and gone to sleep, in their sleeping bags, if they have them. However, the band's ecstatic and literal performance of "Gonna Take You Higher" will get even the most comatose on their feet and dancing. THE WHO are up next at 5am, as ostentatious, bombastic, and loud as befits a set built on their rock opera *Tommy*. Look out for Abbie Hoffman, who will take the stage just after "Pinball Wizard." He will grab a stage mike and deliver a rambling speech about the imprisoned leader of the White Panthers. Pete Townshend, the Who's guitarist, will smack him with the back of his guitar and knock him off the stage.

The dawn will break to the accompaniment of the eerie, eclectic sound of San Francisco's trippy-dippy JEFFERSON AIRPLANE. Guitarist Paul Kantner will recall, "The crowd was into it once they were awake. There were just a lot of people sleeping-bagged out. People were crashing and burning." So do try to wake up before their closing number "White Rabbit" sends you off into a surreal wonderland of half-sleep and dreams.

SUNDAY, AUGUST 17TH

On Sunday morning you may find yourself and your neighbors a little slow to get going. The stage will follow suit, with nothing until 2pm, when JOE COCKER AND THE GREASE BAND will appear. First the band, all tripping on LSD, will be loosening up with a few instrumentals and then the stuttering, shambling Cocker will join them. They will amble through Dylan's "Just Like a Woman" and "I Shall be Released" but hit their peak with a sensational heartfelt take on "With a Little Help From My Friends." You should probably stay right to the end of the set, but be alert: slate-grey storm clouds will be gathering behind the stage. As the applause rings out, the winds and the rain will arrive. They are going to lash you and everyone else for the next hour or so. If you do stick around, you can enjoy the forlorn attempts of the MC and the crowd to turn back the elements chanting, "no rain, no rain." There is also the opportunity for some serious mudslide lunacy at the top of the hill.

The rains will let up after five o'clock. By six the sun will be shining again, and if you are hawk-eyed you will see a small bespectacled figure, pipe in hand, walking up to the microphone. MAX YASGUR'S SPEECH is one of the highlights of the day. The assembled multitudes will feel an immense wave of good vibrations and love as his voice rings out with the words "I think you people have proven something to the world—that half a million kids can get together and have three days of fun and music, and have nothing but fun and music. God bless you." Then it's back to the music, with COUNTRY JOE AND THE FISH rehashing Joe's set from Saturday lunchtime, this time as a band. Just after eight o'clock

REMEMBER WHAT WE SAID ABOUT NOTING JUST WHERE YOU LEFT YOUR CAR? YOU DON'T WANT TO RELIVE THE WHOLE '70S, DO YOU?

the pace will be cranked up by British blues rockers TEN YEARS AFTER, with a pumping version of Sonny Boy Williamson's "Good Morning Little School Girl," and the manic but articulate guitar of Alvin Lee on "I'm Going Home."

You may well be flagging at this point, so it is good to know that the mood is about to become more sedate and reflective. THE BAND, who have recently made their first album without Bob Dylan, put in a strong folk-rock set, notable for "The Weight," which everyone knows from the *Easy Rider* movie. JOHNNY WINTER's blues-rock guitar starts on the midnight hour, and this albino Texan axeman should carry you through to the spacey jazz rock of BLOOD, SWEAT & TEARS at half past one on Monday morning. Stamina will be further rewarded when at three o'clock the recently formed CROSBY, STILLS AND NASH sit gingerly on their high stools and open with an immaculate version of "Suite: Judy Blue Eyes" with soaring, pitch-perfect harmonies and sweet and melancholy melody. After their own acoustic set they will be joined by the enigmatic, bad-tempered but extraordinary NEIL YOUNG, first playing a duet with Stephen Stills before they all play a short electric set including the haunting "Long Time Gone."

Dawn is coming. You have been lulled into an ethereal dream by the plaintive harmonies of Crosby, Stills and Nash. You may be simply mud-caked, rain-battered, and exhausted, but there are still five hours to go, and the PAUL BUTTERFIELD BLUES BAND is going to get you there. Starting at six in the morning they will be serving up a set of uptempo Chicago white-boy blues, with talking guitars and wailing harmonica. If that isn't enough musical caffeine for you, SHA-NA-NA's maniacal set of retro pop classics should be. Their all-singing, all-dancing covers of "Jailhouse Rock" and "At the Hop" are unfeasibly upbeat for eight o'clock in the morning.

There is then an hour's pause—for whatever you want, man—before the finale. At just after 9am JIMI HENDRIX and a five-piece band will open with "Getting My Heart Back Together Again." Two hours of wild psychedelic rock later, the amphitheater will feel like an abandoned battlefield. The crowd will have shrunk to fewer than 50,000, leaving a landscape of abandoned possessions, piles of rubbish, and smouldering fires. In the midst of this, Hendrix will announce that he is "just jamming" before unleashing his extraordinary version of the "Star-Spangled Banner" and closing with a spine-tingling rendition of "Hey Joe." You are witnessing musical history.

PART TWO

Moments That Made History

The Boston Tea Party

DECEMBER 15–16, 1773 ✵ NEW ENGLAND

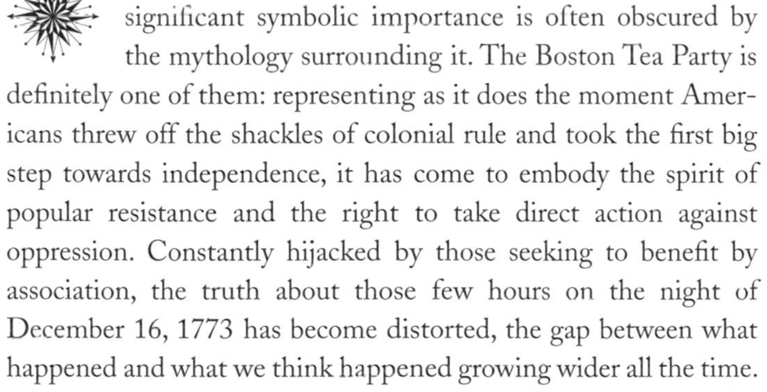

ANY HISTORIC EVENT THAT HAS ACQUIRED significant symbolic importance is often obscured by the mythology surrounding it. The Boston Tea Party is definitely one of them: representing as it does the moment Americans threw off the shackles of colonial rule and took the first big step towards independence, it has come to embody the spirit of popular resistance and the right to take direct action against oppression. Constantly hijacked by those seeking to benefit by association, the truth about those few hours on the night of December 16, 1773 has become distorted, the gap between what happened and what we think happened growing wider all the time.

Even the name is misleading; it wasn't called the Boston Tea Party until the 1820s and, as you will see, it was not much of a "party." Moreover, due to the fact that the participants swore an oath of secrecy—which they all observed—we don't know with any certainty who was responsible for dumping the tea. Only after they were all dead did their "names" start to appear in print.

In 1835, over sixty years later, a "witness" produced a list of 58 individuals he claimed were on the ships. Not long after, another list was culled from interviews with Bostonians. There is no way of being 100 per cent sure of their authenticity.

This lack of clarity presents the time traveler with a unique opportunity. You can actually *be* one of those anonymous folk heroes—not merely a witness, but a part of the legend. Without disturbing the time-space continuum. During thirty-six hours in Boston, you will sample the town's thriving tavern culture and its busy streets thronged with traders of all kinds, experience the intense buildup to the Party, and join its citizens on the waterfront as they consign the British tea to a watery grave.

BRIEFING: REVENUE ACTS AND MASSACRE

Victory in the Seven Years' War (1756–63) made Britain the world's dominant imperial power but left its government £145 million in debt. In 1763, hoping to recover some of this by raising higher revenues from its American colonies, Parliament enacted *the* first of a series of laws—the *AMERICAN REVENUE ACT*, increasing duty on imports of textiles, coffee, indigo, and foreign goods—that would inspire cries for "no taxation without representation." Next came *the STAMP ACT* of 1765, which hit merchants, publishers, and lawyers especially hard.

Furious, the people of Boston went on the rampage, ransacking the sumptuous homes of the city's elite, whose wealth and influence rested on their relations with the British. Mass demonstrations followed. Stung by this reaction, Parliament repealed the Stamp Act within six months. However, in 1767 it introduced the *TOWNSHEND ACTS*, which increased duties on a range of goods, including tea, with similar results.

The situation in Boston became so serious that a detachment of soldiers was sent to regain control, a move that went spectacularly wrong on February 23, 1770 when they opened fire on a crowd that had been taunting them, killing five people. Fearing for their safety after what became known as the BOSTON MASSACRE, the troops retired to the sanctuary of Fort William on Castle Island.

This period of sustained resistance shaped the radical movement in Boston; leaders emerged, 300 citizens joined the SONS OF LIBERTY, and alliances were forged between the town's middle and lower classes based on networks of interdependence that served to isolate the state governor, Thomas Hutchinson, and his small, nepotistic cabal of allies.

However, aside from a sustained boycott against British tea, the early 1770s were relatively calm compared to the upheavals of the previous decade. Then, in May 1773, Parliament introduced the *Tea Act*, hoping to help the ailing East India Company, which had a monopoly on the tea trade. Chartered in 1600 by Queen Elizabeth I, the Company was on the verge of bankruptcy, with £17 million of unsold tea sitting in its British warehouses. Though the Act actually lowered duty on tea to try to undercut the thriving black market in smuggled Dutch tea, it was met with predictable fury by the Boston radicals.

The Dartmouth was the first vessel operating under the terms of the Act to arrive in Boston Harbour; it was half-owned by Francis Rotch, a 23-year-old Quaker merchant from Nantucket, who was also responsible for its cargo. On November 30, it docked at Griffin's Wharf. By December 1, all the cargo had been unloaded except the tea. Two days later, a smaller ship, the *Eleanor*, pitched up, followed by the *Beaver*, smaller still, on December 7.

Once the *Dartmouth* had been officially registered by Customs men, it had twenty days to unload and pay the necessary duty or all its goods would be seized and the British state would reap the benefits. With the deadline set—midnight on December 16—and the clock ticking, tension began to build inexorably. On December 14, at

PAUL REVERE'S 1768 ENGRAVING OF BRITISH TROOPS ARRIVING IN BOSTON—AN IMAGE THAT INCENSED AMERICA.

the largest public meeting ever held in Boston, Rotch was put under severe pressure to take his ship and sail away; but if he did he was certain to lose his investment. So he prevaricated as best he could.

With members of the SONS OF LIBERTY mounting a 24-hour guard to prevent the tea coming ashore, its fate was still undecided. Would the people of Boston act decisively and defy the Crown? And, if so, would the British Navy, represented by the 64-gun HMS *Captain* and the frigates *Active* and *Kingfisher*, intervene to stop them?

THE TRIP

You will arrive in Boston in the early afternoon of Wednesday, December 15, by the LIBERTY TREE on High Street near Hanover Square. This huge elm tree was planted in 1642 to mark the beginning of Parliament's war against the "tyrant" Charles I. It is the focal point of rebellion, site of public meetings, and the place to hang effigies, proclamations, and calls to action.

You will be dressed as a member of the labouring classes (they will form the majority of those at the "party") who has come to Boston from rural Massachusetts. Men will be wearing worn black leather shoes, leggings up to your breeches, a heavy woollen coat down to your knees, a thick HUNTING SHIRT, and a folded beaver felt hat. Underneath a plain cloak, jacket, and petticoat, female travelers will be squeezed into a STAY, a cone-shaped, corset-like garment designed to support your bosom and provide good posture, which is very popular with women employed in manual work. To help stave off the cold, you will have a mob cap and fingerless mittens.

You have a bed reserved at THE WHITE HORSE TAVERN, owned by Joseph Morton and situated on Newbury Street near Frog Lane, recognisable by its signage, which features a white charger. Boston has over 150 taverns—from high-class joints to disreputable dives—and the White Horse is one of the best. Entering from the street, you will step into the public bar (tap room) with its beamed ceilings, hard white-oak-sanded floors, and fireplace full of blazing logs. The room will already be full of tobacco smoke—clay pipes are available for rent—and punters will be engaging in games of cards and skittles, using a small ball known as the cheese. You will notice rows of barrels near the waist-high bar, salted cod-fish snacks hanging by it, and a selection of religious reading material for customers to peruse.

As you make your way to the back of the room, you will see a small box nailed to the wall in which you can leave TIPS for the servants; it has a tiny opening and the words *To Insure Promptness* inscribed on it. Exit through the door and go upstairs where you will find your bed in a room with several others—sharing is the norm. The facilities are rudimentary to say the least: nowhere to wash and toilets that are not much more than a covered hole in the ground.

BOSTON

Boston, its skyline dominated by steeples and masts, pokes out into the Atlantic and the only thing preventing it from being an island is the narrow strip—BOSTON NECK—connecting it to the mainland. The town is divided into two: the more upmarket NORTH END and the rougher SOUTH END. Blessed with a very long shoreline, crowded with warehouses and shipbuilders, and its proximity to the New England fishing and whaling grounds—source of ever versatile whale oil and whale bone—the city is a mercantile trading hub bustling with entrepreneurial activity. This is reflected by the variety of retail outlets (over 500) that will compete for your attention as you wander the streets, the air reeking of sea salt, horse manure, and trash.

You will pass grocery stores with staples like sugar and corn, shops packed with different sorts of cloth (calico, chintz, muslin, cotton, buckram, silk), haberdashers, dressmakers, hardware stores, leather goods shops, and even prototype DEPARTMENT STORES housed in warehouses that stock all manner of items from

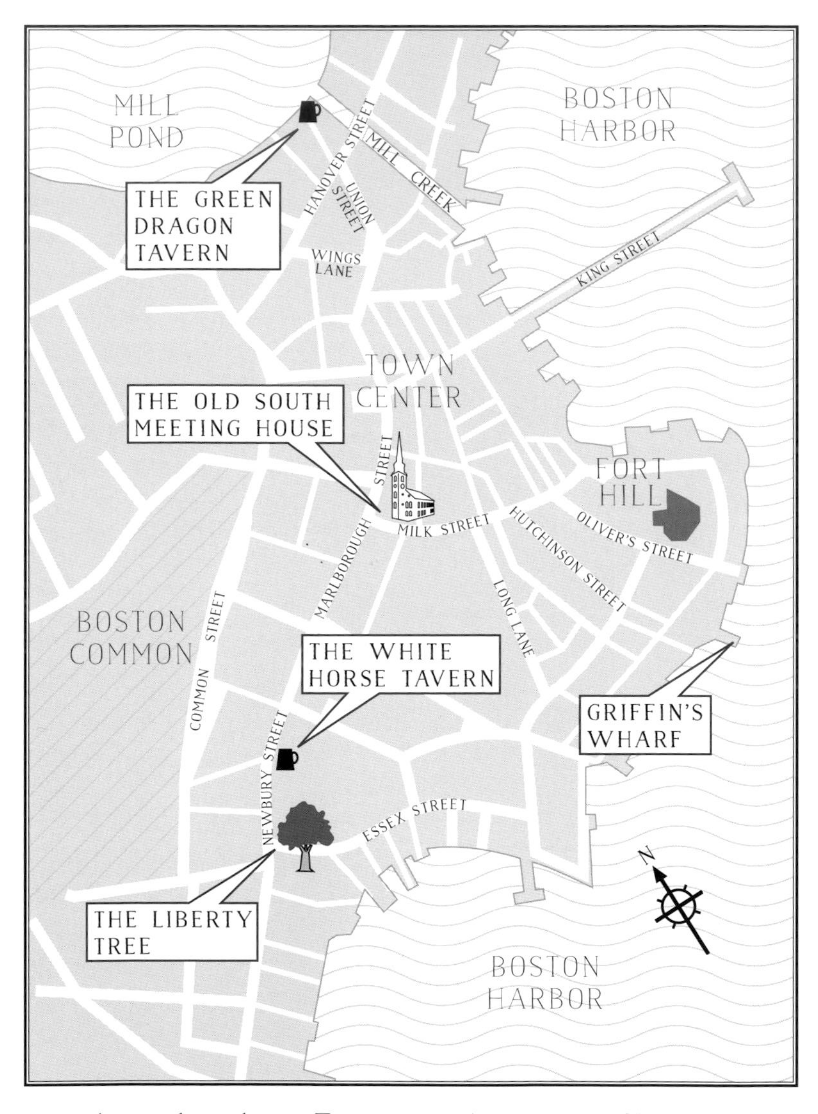

cutlery to hourglasses. For a more sobering view of buying and selling, there are SLAVE MARKETS in King Street.

You may also be surprised by the number of BOOKSTORES; Boston has a very high literacy rate thanks to colony-wide compulsory education. Every citizen is expected to be able at least to read the Bible. As a result, the book trade is booming, and many of those involved have moved into printing as well. Other than religious tracts, typical booksellers, such as JAMES FOSTER CONDY, offer poetry, philosophy, history, and political pamphlets, while THOMAS HANCOCK also stocks stationery (paper, quills, sealing wax, ink-horns). For LOCAL NEWS, pick up the *Boston Gazette* or the weekly *Massachusetts Spy*.

If you find yourself getting a bit peckish, then try the fresh OYSTERS available from street-sellers. For a hot beverage and somewhere to put your feet up, there are dozens of COFFEE-HOUSES serving smuggled Dutch coffee and cocoa. However, do avoid the ROYAL COFFEEHOUSE as it is largely frequented by those loyal to the British Crown and the officials that represent it.

EATING AND DRINKING: TAVERN CULTURE

DINNER is served at the WHITE HORSE at 7pm, though given the range of taverns to choose from don't feel obliged to eat here; the BUNCH-OF-GRAPES on King Street is an excellent alternative. At either venue you will be treated to a good meal: broth to start, followed by seafood in the form of lobsters, eels, or salmon, baked or fried; a main of roast meat—veal, pork, beef, or turkey—served with a potpourri of root vegetables; a side salad of thinly sliced cabbage; bread and cheese; and a dessert to finish, either fresh fruit, Sippet Pudding (bread, raisins, eggs, and sugar), or Blown Almonds.

There are French and Spanish WINES available, but the citizens of Boston prefer hard CIDER (from apples or pears), BEER, or RUM. Beer—of which there are both locally brewed and exotic herb-infused brands such as *Spruce* and *Birch*—is such a fixture of life that a whole vernacular of slang expressions has evolved to describe the experience of drinking it.

Served in pint-size pewter BEERPOTS (tankards) or in brown- or green-coloured bottles to prevent sunlight rotting the contents so they smell bad (skanky), the strongest beer is known as comfort for the poor. Whatever you do, avoid a beer known as old trousers, a foul mix of left-over dregs. Any beer you might leave temporarily to one side which is then drunk by someone else is called a hamster, while drinking slowly is known as tippling. If you get drunk, you will be referred to as an ale knight or a toss pot who has allowed himself to get carried away by ale passion.

If beer is not your thing, then why not try some DARK RUM—45 per cent alcohol and 90 per cent proof—from one of the twenty local distilleries based in the Essex and South Street areas. Made mostly from molasses smuggled from the French and Spanish Caribbean sugar islands, which are a third cheaper than the British equivalent, the standard measure is called a DRAM. You can imbibe it neat or in a primitive cocktail mixed in a large earthenware pitcher (muller). Favourites include black strap, a very strong punch of rum and molasses; calibogus, an unsweetened rum and beer concoction; the ever-popular flip, a mix of rum, warmed beer, brown ale, eggs, and sugar that is stirred together by a long-handled wooden spoon (bar spoon); a stone fence, which is rum and hard cider; or a rattle skull that combines dark beer, rum, lime juice, and nutmeg.

THE GREEN DRAGON TAVERN

Aside from being social and cultural centres, Boston's taverns are also at the heart of radical politics and are where opposition to British policy is debated and planned; tonight all conversation will be focused on the looming tea deadline. If you want to be close to the decision makers, then head over to the GREEN DRAGON TAVERN on the corner of Hanover Street and Union Street in the North End of town. Bought by the St Andrew's Lodge of Freemasons in 1766, its upstairs room is the meeting place for the town's most prominent rebels, many of whom are Harvard graduates.

Take a pew in the public bar and you will see SAMUEL ADAMS, head of the Committee of Correspondence and de facto leader of the SONS OF LIBERTY; PAUL REVERE, a silversmith of Huguenot

A COLONIAL POLITICAL CARTOON FROM 1767 CHARACTERIZES THE INTRODUCTON OF TAX ON TEA AS "WORSE THAN A BAYONET" TO "VEX THE SOULS OF THE MEN OF BOSTON TOWN"

descent who, on December 17, will ride to Philadelphia to spread news of the Tea Party; JOHN HANCOCK, local benefactor, Grandmaster of the Lodge, and the wealthiest merchant to join the rebel cause; JAMES OTIS JR, a lawyer, sadly on the verge of insanity due to head injuries inflicted on him by an enraged Customs Commissioner; WILLIAM MOLYNEUX, a merchant running his own "mob" of craftsmen and small retailers; and EBENEZER MACKINTOSH, a debt-ridden shoemaker from South End who commands around 150 waterfront toughs. All will come in and head upstairs to thrash out the agenda for tomorrow's protests.

THURSDAY, DECEMBER 16TH

The White Horse serves a hearty breakfast at 9am that will help combat your hangover: herring, smoked or pickled meats such as ham or gammon, toast, bacon, and eggs. A gentle hair of the dog comes in the form of SMALL BEER (weak beer—only one per cent alcohol), or you can have strong sweet black coffee if you prefer.

Once you're done and, despite the biting cold rain that will persist all day, we suggest you walk off your heavy meal and make your way to BOSTON COMMON for a stroll round America's oldest public park. Ranging over 50 acres and home to a considerable number of cows, it was established in 1634. The grounds are crisscrossed by walkways and contain several perimeter promenades such as the Tremont Mall and a bandstand, recently added courtesy of John Hancock.

THE OLD SOUTH MEETING HOUSE:

By 10am, you will need to be at the OLD SOUTH MEETING HOUSE (the largest church building in town, with a 183-foot-high steeple completed in 1729) on the corner of Washington Street. Here you will join around 5,000 Bostonians (out of a total population of 16,000) who are gathering to decide what to do about the tea. Despite the heat emanating from all these bodies packed together, the hall will still be freezing, as there is no fireplace: you may find yourself jealously eyeing the footwarmers and portable stools that some people have brought with them.

The first order of the day will be yet another cross-examination of the hapless Rotch, culminating in him agreeing to seek out Governor Hutchinson and ask him to allow the ships to leave the harbour with their cargo unloaded and return unmolested to the UK. As soon as he departs, the meeting is adjourned until 3pm and the senior radicals retire to an adjoining room to discuss tactics.

At this point we suggest you return to the White Horse for lunch, after which you will have to decide whether or not you have the necessary physical strength, fitness, and stamina to endure the extremely hard and intense labour that lies ahead for those engaged in destroying the cargoes of tea. Remember: your colleagues will not tolerate any slacking off or weak team players; however tempted you are, do not bite off more than you can chew. If, however, you feel both capable and willing, then wait in the public bar until word filters through about the planned action, as it will, and then head up to your room to don your disguise.

ALL THE TEA IN CHINA

All the TEA that is about to be hurled into the harbour comes from China. The majority of it is a BLACK TEA called *Bohea* from Jiangxi province; cheap and long-lasting, with an earthy aroma and strong flavour, it comes from leaves that have been fully fermented. There will also be a certain amount of more refined black teas like *souchong* and *kung fu* ("labour requiring skill and effort").

Alongside these varieties will be some GREEN TEAS, primarily a brand called *Hyson* from Hi-Thsun plus a scattering of the finest type, *Singlo*, from Anhui province; more delicate and therefore more expensive, green tea leaves have been heated and steamed immediately after picking to prevent oxidation and create a sweet, fresh, and mellow taste.

Derived from the plant *Camellia sinensis*, an evergreen shrub with firm leaves containing caffeine and antibacterial chemicals that can grow up to 12 feet high and live for 100 years, Chinese tea is hand-picked by women—a back-breaking task—and then the leaves are sorted from the stalks, cured through heating, and rolled out, usually on the same day. Though there are three or four harvests a year, the main season runs from mid-spring to early summer.

The processed leaves are collected by tea peddlers, then sold on to wholesalers, who sort and pack, then on to merchants who transport their wares downriver to Canton, arriving late November/early December, where the tea is sold to EAST INDIA COMPANY representatives. It is placed in wooden tea chests lined with lead to prevent contamination, and packed together with strong and flexible rattan (bamboo) matting before being loaded onto waiting ships. From there it will take six months to reach London, where it will be auctioned off by the Company to merchants like poor old Francis Rotch.

MOHAWKS

You and your comrades will be dressed up as Native Americans from the MOHAWK tribe. Part of the Iroquois Confederacy, the Mohawks are known as the KEEPERS OF THE EASTERN DOOR and are feared and respected for their prowess in battle and military organisation. Adopting this disguise serves several purposes; to make you difficult to recognise, even though it remains obvious that you are not actually Mohawks (after all, there are only thirty-seven Native Americans in Boston); to ensure deniability and provide a convenient scapegoat; and to exploit the Mohawks' image as representatives of pre-colonial Americana and the mystique surrounding their unfettered way of life.

The authenticity and elaborateness of the disguises will reflect the status of the individual wearing them. Your eighteen leaders will wear the most convincing outfits, with headdresses, faces darkened by charcoal paint, robes, cloaks, and Mohawk-style shawls, which will completely conceal their regular clothes. They will even issue orders in Native American dialects.

The quality of costume will then deteriorate as they pass down the ranks; some will use lampblack or grease to cover their faces and make do with woollen hats and old frocks. You will be provided with a couple of blankets to drape round you and soot to blacken your features. You will also have an axe and a hatchet, which you will hang from your belt. Once you are ready, head for the GREEN DRAGON where other Mohawks will already be congregating. Together you will proceed to Milk Street, not far from the OLD SOUTH MEETING HOUSE where matters will be coming to a head.

ROTCH'S RETURN

If you decided tonight's work is beyond you, then be back at the OLD SOUTH MEETING HOUSE by 3pm. You can listen to a series of speeches while you wait for Rotch's return. You will hear JOSIAH QUINCY, sadly suffering from TB, urge you to resist, but remember that the consequences will be severe if you go too far; and then HARRISON GRAY, the colony's treasurer, will answer him.

At 5:45pm, with candles now flickering round the hall, Rotch reappears, having completed the 14-mile round trip to see Hutchinson at his country house in Milton (he'd abandoned Boston some time ago). Rotch will inform you that Hutchinson refused his request. His words are greeted by uproar, voices yell out, "Boston harbour a teapot tonight," and from the podium John Hancock sombrely declares open season: "Let every man do what is right in his own eyes."

Inflamed, the crowd begins to rumble into motion. Suddenly, you will hear spine-chillingly accurate Mohawk war-whoops coming from outside, which will trigger a surge towards the exit as you realise that these bloodcurdling cries signal the start of something. Streaming out of the building, you will see a core group of finely attired Mohawks carrying torches whose stony silence is infectious; soon everybody will fall quiet.

Without further ceremony, the Mohawks will lead you down Milk Street. Their numbers keep swelling as they make a sharp right onto the waterfront and arrive at the wharves and warehouses by Fort Hill. It is here that the crowd will stop and become spectators while the Mohawks carry on to GRIFFIN'S WHARF and the tea clippers.

THE PARTY

At 7pm, with a bright moon shining in the night sky to illuminate your activities, and the British warships stationary at their posts some 400 yards away, you and between 80 and 100 others will clamber on board the *Dartmouth*, then the *Eleanor*, and finally the *Beaver*. Your companions are mostly under thirty years of age and consist of journeymen, apprentices, dock labourers, navvies, some skilled craftsmen and artisans—carpenters, builders, coopers, leatherworkers—and a handful of professionals.

The operation will run like clockwork, a disciplined well-oiled machine. Marshalled by self-appointed captains and boatswains, flanked by sentries armed with pistols, your first task will be to order the ship's crew to open the hatches to the hold and hand over the hoisting tackle that you will use to haul up the tea chests—each weighing 400 pounds—decorated with black-and-red Chinese motifs.

Once on deck, axes will be used to split open the chests, then the contents chucked over the side. Save for the sound of splintering wood, the deck creaking underfoot, and the splash of tea hitting the water, you will toil in total silence. You are also under strict orders not to damage any ship's property or steal any of the merchandise; on the *Beaver*, one CHARLES CONNER will be caught stuffing his pockets with tea and suffer the indignity of being stripped, coated with mud, and beaten up.

You will notice the water around the ships becoming distinctly murky as more and more tea is dumped into it. Because the tide is very low and the ships are riding in only a few feet of water, the tea will pile up on the seabed, clogging up the bay. Some spectators will grab a rowing boat and try to salvage some of the valuable leaves floating on the surface, but they are intercepted before they even get in the water. To prevent any others from doing the same, some of you will be asked to wade in and break up the thick conglomerations of tea by stirring it into shallow water or stamping it into the mud.

Boston Boys throwing tea into the harbour

THE MOHAWKS GET DOWN TO PARTYING.

After two hours, all of the 342 tea chests—amounting to 46 tons of tea, with a total value of £9,654 (about $1.5m in today's money)—will have been tossed into the harbour. Only one, *The Robinson Tea Chest*, survives intact and will be discovered the next day by the teenager John Robinson, buried in the sand along the shores of the Atlantic.

IN THE CROWD

From its vantage point, the 2,000-strong crowd will only be able to see a mass of shadowy figures moving busily and efficiently around the ships. Despite the tension and excitement of the occasion, they wait patiently, hold their ground, and maintain a hushed reverential silence throughout. No one cries out. No one cheers. No one breaks into song. It's as if you are witnessing a well-rehearsed, perfectly coordinated performance, the spectators instinctively aware that they are watching history being made.

Job done, the crowd will swiftly melt away. The Mohawks will line up together in military formation, a fife will strike up and they will march back up the wharf and onto Hutchinson Street where they will quietly disperse. By 10pm, the streets are completely deserted, and the only evidence of your evening's work is the tons of tea floating in Boston Harbour.

DEPARTURE

Before leaving the dockside, check that there aren't any stray tea leaves on your person—many of your fellow wreckers will get home to find their shoes and coat pockets full of tea and have to resist the urge to boil some water and make a refreshing brew. If you do find any remnants, get rid of them immediately, and then return to the LIBERTY TREE by midnight for departure.

The Women's March on Versailles

OCTOBER 4–5, 1789 ✵ PARIS/VERSAILLES

THE FRENCH REVOLUTION WAS ONE OF the greatest upheavals in history: its defining feature was chaos; its outcome anything but inevitable. At various crucial moments the people of Paris made decisive interventions, each time pushing the Revolution in a more radical direction. THE WOMEN'S MARCH ON VERSAILLES established the template for these mini-insurrections. Its display of strength emboldened the people, giving them the confidence to take control of events and bend them to their will.

Over forty-eight hours a group of tough working-class Parisian women—their contemporaries called them Amazons—ransacked the PALACE AT VERSAILLES and forced KING LOUIS XVI and QUEEN MARIE ANTOINETTE to up sticks and relocate to Paris, where they became virtual prisoners. Their stunningly opulent palace was left abandoned, its sumptuously decorated rooms and mirrored halls deserted, the doors shut, the gates locked, the glorious gardens patrolled by armed guards. A few years later, both the king and queen would go to the guillotine, a journey that began on the day the women came to town.

On this trip, you will spend a NIGHT OUT IN PARIS, enjoying the pleasure gardens of the PALAIS ROYAL, where the whole city goes to play. Next day (October 5, 1789), you will join the women on their march for bread and storm the Palace at Versailles, returning in triumph to Paris to party a second night away.

BRIEFING: REVOLUTION AND THE "AUSTRIAN WHORE"

Earlier in the year (July 14th), Paris rose up to LIBERATE THE BASTILLE, the most loathed and notorious prison in France. It was the symbol of everything rotten about the *ancien régime*—a system that the delegates of the newly convened NATIONAL ASSEMBLY are determined to dismantle—and a cause for much celebration. However, the euphoria has not lasted for long. While the National Assembly has spent August consigning feudalism to the trashcan of history and framing the DECLARATION OF THE RIGHTS OF MAN, the citizens of Paris increasingly fear a counter-attack from the forces of reaction, either a direct intervention by a foreign power or an aristocratic plot. Adding to the mounting tension are severe bread shortages; though this summer France has had the first decent harvest in years, a drought has meant that millers did not have enough water to grind the corn. The price of a standard four-pound loaf has risen over the year by 60 percent. Throughout September, bakers have been attacked for overcharging and grain carts have been randomly seized.

By October, Paris is on a knife edge. And on October 2nd there is a perfect provocation when a lavish banquet, attended by Marie Antoinette, is held in honour of the king's Flanders Regiment (the Black Musketeers), who have been called to Versailles to bolster the royal defences. Over the course of a debauched evening, the soldiers sing monarchist songs; they wear hats emblazoned with cockades in black (the queen's color) or white (the king's) rather than the TRICOLOR that is an essential item for all true supporters of

MARIE ANTOINETTE PLAYING AT BEING A PEASANT.

the Revolution. Worse still, they allegedly curse and trample on the Revolution's tricolor cockade. When news of these outrages reaches Paris, it is swiftly inflated. Indeed, the gossip is that Marie Antoinette participated in an orgy with members of the regiment. These rumours lead inevitably to the Women's March on Versailles.

On the march, you may be shocked by the women's intense hatred of MARIE ANTOINETTE and the savagery of the abuse they hurl at her, often sexual in nature. Married at fifteen, queen at nineteen, Marie Antoinette secured a bad reputation by dressing inappropriately and attending Parisian theaters and decadent social gatherings without her husband. In 1777, *Anandria*, the first of what became a steady stream of pornographic novels about her, was published. It included generic elements—lesbianism, nymphomania, and masturbation—that would be recycled in songs, satirical verses, and other books. The most widely read is *Essai Historique sur la Vie de Marie Antoinette*, first published in 1781, reprinted in 1783, and then updated on a yearly basis until her death a decade later.

More firmly based in fact is the belief that the queen is as profligate with money as she is in the bedroom. While ordinary people starve, she spends like there is no tomorrow and has a particular weakness for diamonds. This toxic combination of material and moral depravity, both real and imagined, will account for much of what you see and hear over the next few days. If, however, you are expecting to hear her say, "Let them eat cake," or something similar, you have been woefully misled; she never said any such thing. As you will see, it would have been most unwise.

THE TRIP

You will arrive at 3:30pm on SUNDAY, OCTOBER 4, 1789, in the courtyard of the HÔTEL DE SOUBISE, 60 Rue des Francs-Bourgeois, in the heart of Paris. The Soubise is what is known as a *hôtel particulier*—a very posh B&B—one of the grand mansions of the super-rich that stand empty most of the time, as their owners spend no more than a few days a year in Paris. Our choice belongs to the Prince de Soubise—who is characteristically absent—and has dozens of rooms decorated in a rococo style. It has a small army of servants and is lavishly appointed with exquisite furnishings, fine art on the walls, marbled baths, bidets, and king-size beds. Both sexes will have their own upholstered POTTY CHAIRS with a hole in the seat and a chamberpot underneath, the men's receptacle purely functional, the women's an ornamented bourdaloue, which looks like a gravyboat. The servants will be on hand to empty them.

Your clothes for your night out in Paris will be typical of the city's professional middle classes. Men will be wearing a wig, dark-green jacket (open at the front with tails), white leggings, calf-length leather boots, and a top hat. Women will have their hair up in a bunch, a shawl and a relatively plain low-cut dress with a voluminous floor-length skirt and lace frills. Men may carry a silver-topped cane, women a parasol.

SUNDAY NIGHT AT THE PALAIS ROYAL

THE PALAIS ROYAL is the place to be after nightfall. Open to all and sundry, it belongs to the Duc d'Orléans, who inherited it in 1776 and decided to transform it into a pleasure palace. Its three-tiered arcades run in galleries round the perimeter of the site and

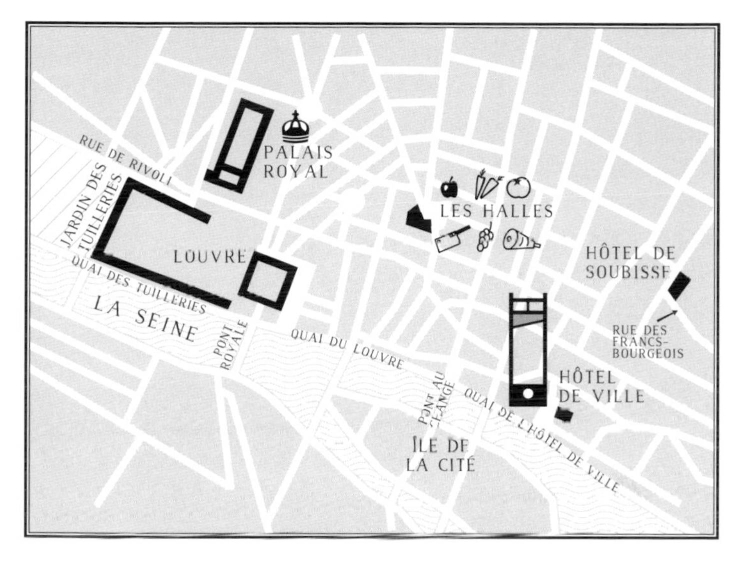

contain sixty pavilions, all enclosed by trees, which line a large promenade. Though the SHOPS will be closed—it is Sunday, after all—you can gaze at the fashionable clothes, designer wigs, and exquisite lace garments displayed in their windows.

On the PROMENADE, there will be hawkers selling all manner of items and you will be able to select from the plethora of publications that have been released since the Revolution did away with censorship; newspapers like *Révolutions de Paris*, with its racy prose, eyewitness reporting, and captioned illustrations, or Marat's incendiary *L'Ami du Peuple*, which urges readers to "sweep away the corrupt, the royal pensioners, and the devious aristocrats." Also on offer will be SATIRICAL PAMPHLETS, political tracts, and ballad sheets. Cheap FIREWORKS—for twelve sous you can buy squibs, rockets, and serpents—are on sale everywhere and provide a constant backdrop of noise and color, adding to the carnival atmosphere.

As you wander around, you will see LIMONADIERS (lemonade sellers) plying their wares, and we suggest you try some: *limonade* is France's favourite soft drink and a source of national pride, considered the best in Europe. This is not surprising, given the

CAMILLE DESMOULINS ADDRESSING A CROWD AT THE PALAIS ROYAL, OUTSIDE THE CAFÉ DE FOY.

care that goes into its preparation. The yellow rind from three or four lemons is sprinkled into a jug filled with clean spring water, covered, and left to stand for a few hours. Next, the juice from the lemons is added and, after thirty minutes' waiting time, filtered eight times through a coarse linen cloth. After adding sugar the whole mixture is run through the cloth two more times, resulting in an extremely refreshing and thirst-quenching beverage. After drinking a glass or two, you may need to empty your bladder; if so, there are public latrines—enclosed rooms with rows of wooden seats and a lunette (cover) over each hole.

For contemporary popular entertainment, the Palais offers the THÉÂTRE BEAUJOLAIS, featuring three-foot-tall marionettes and child actors, and the VARIÉTÉS AMUSANTES, which stages music-hall-style farces and melodramas to packed houses. Novelty attractions—MAGIC LANTERN and SHADOW-LIGHT SHOWS—can be found in the wooden galley running along the Palais, alongside a kind of FREAK SHOW; one of the biggest draws is Paul Butterboldt, the FOUR-HUNDRED-POUND MAN; also look out for a DEFROCKED PRIEST singing obscene songs while strumming his guitar.

A must-see for all visitors are the WAX-WORKS DISPLAYS at Salon No. 7, mounted by Philippe Curtius, a pioneer of this form of spectacle and owner of Paris's first waxwork museum, *Le Grand Couvert*. At the Palais you will pay twelve sous to sit up close, or two sous at the back. In front of you will be waxwork tableaux featuring various royal families, cultural and intellectual giants like Voltaire, and military heroes.

For those of a sporting disposition, there are BILLIARD TABLES and GAMBLING DENS on the upper floors. After 10pm, flocks of PROSTITUTES will gather by a colonnaded area known as the Camp des Tartares.

EATING AND DRINKING

There is no need to leave the Palais for dinner. Its RESTAURANTS are the finest in Paris. The best, and most expensive, is the CAFÉ DE CHARTRES, where you can enjoy *plats bourgeois* (the general term for haute cuisine) such as fried mutton feet, rabbit steaks with cucumber, roast veal in pastry, and vine-leaf fritters. Slightly cheaper, but no less excellent, is LES TROIS FRÈRES PROVENCAUX, run by three brothers from Marseille who serve regional specialities: *bouillabaisse* (fish soup) and *brandade* (puréed salt cod). Normally with your meal you'd expect a top-quality *pain mollet* (soft bread) like the elaborate *pain de fantaisie*; however, in times of need, when BREAD is scarce, it is customary not to bake such fancy loaves, and the owners will respect this tradition; to ignore it is to invite trouble.

For coffee, a dessert of sorbet or ice cream, and a liqueur to finish, the Palais has plenty of CAFÉS to choose from. The current hotspot is the CAFÉ DE FOY, frequented by leading revolutionaries like Georges Danton and Camille Desmoulins. The COFFEE comes from Santo Domingo in the Caribbean and you can have it strong and black, or with milk (*café au lait*);

you can even get a *mocha*. Take a seat at one of the outside tables under the chestnut trees and you will be surrounded by all manner of Parisians passionately debating politics. Games of CHESS are also available at a number of tables: chess is a common feature of café culture, and standards are high, competition fierce. So if you are an enthusiast, pull up a chair and test your moves against the best in revolutionary Paris. For a more refined atmosphere, visit the CAFÉ DE LA RÉGENCE, with its marbled tables, mirrored walls, and chandeliers.

FRENCH BREAD

The role of BREAD in the events you are about to witness is critical—and symptomatic of its importance to eighteenth-century French society. Available in a range of shapes and sizes, from round to long and thin (don't call them *baguettes*, as the word doesn't appear until 1920), their unique flavour and texture is largely due to their ingredients: salt, either grey from the salt marshes or white from the Normandy seaside; water drawn from rivers, fountains, wells, or directly from rainfall; butter; sometimes milk and sugar; occasionally egg, for a glazed effect. White stone-ground flour is sifted through cloth to create three main types of dough: *ferme* (hard), *molle* (soft), and *bâtard* (bastard), which is somewhere in between the two and is overtaking *ferme* as the standard variety.

Bread also denotes social class. The elite will mostly eat *pain mollet* (soft, crusty, savoury bread); the bourgeoisie, *bâtard* breads like the aptly named *pain bourgeois*; and the vast majority, *pain du commun* (hard, crustless bread). Over the course of this trip, you will sample a range of these loaves and get a taste of what the Women's March is all about.

LOWER ORDERS

If you want to be better acquainted with the lives of the working women and men who will lead you to Versailles tomorrow, we suggest you walk the short distance to the less-affluent neighborhoods near LES HALLES (Paris's main market) and go into any busy TAVERN, where you will be able to get reasonable wine from Bordeaux and Burgundy, and a hearty bowl of soup. The bread, if there is any, will either be *pain à soupe*, which is a very flat loaf made entirely from crusts and *bâtard* dough, or a *pain du commun* (hard) like the local staple, *gros pain de Paris*. For a more down-market experience, check out one of the many street-corner WINE SHOPS, where you can buy a bottle to take away, or stay and drink standing up (there are no chairs) and mix with labourers on a daily wage—market porters, chimney sweeps, coachmen—as well as criminals and hookers.

In all these venues, you will hear the clientele speaking POISSARD, the favoured slang of the market women and fishwives. Derived from the word *poix* (pitch), it is a crude but vibrant patois featuring heavy elision, fractured syntax and grammar, plus rudimentary rhyme, perfect for jokes and ribald songs. It has quickly become fashionable, taken up by the bourgeoisie and the aristos, who get a kick out of dropping it into polite conversation.

MONDAY, OCTOBER 5, 1789

After your late night—Paris does not sleep until about 4am—you will need to ready for BREAKFAST by 7am, dressed in your PROLETARIAN OUTFIT. Women should be wearing a white bonnet, loose white blouse and a white apron over a billowing pleated skirt. Men should wear a red cap, white shirt with a sleeveless brown leather jacket, and blue *culottes* (knee-breeches). Both sexes will have a TRICOLOR COCKADE, best worn as a sash, either wrapped round your waist or across your torso.

At your local patisserie try a *pain à café*, which is a rich breakfast roll. Obviously it goes well with coffee, but we recommend you order a *chocolat chaud* instead: either a dark-chocolate brew—vanilla and sugar mixed with cocoa paste—or a creamier version with milk. Since its arrival in France in the seventeenth century, HOT CHOCOLATE has been growing in popularity, and these days it's all the rage—Marie Antoinette has her very own *chocolatier* working on new recipes—and spices like cinnamon, nutmeg, and clove are common additions. If you are fed up with bread, and would like something different, try a *choux* bun; made by mixing butter, water, flour, and eggs to create a light choux dough, this type of bun was recently invented by the *pâtissier* Avice.

While you are eating you will hear the tocsin sound from the Church of Sainte-Marguérite, followed by the BELLS of dozens of others summoning the citizens of Paris onto the streets to answer the call of a young woman who is banging a drum and yelling, "When will we have bread?" As their numbers swell to nearly 7,000, THE WOMEN—mostly fishwives, shopkeepers, market traders, and pedlars from the Faubourg Saint-Antoine, armed with cudgels, knives, and sticks—are joined by men from the surrounding districts. Together the angry mob will proceed to the HÔTEL DE VILLE, Paris's town hall and the focal point of mass protest.

AT THE HÔTEL DE VILLE

You should be in the PLACE DE GRÈVE, outside the Hôtel de Ville, by 8am. On arrival, you will see the women, and some men, breaking down its locked doors. The more adventurous may want to follow them as they burst in, tearing up documents and ledgers as they search for weapons; however, this is not an occasion for wholesale looting—a wad of a hundred 1,000-livres notes goes missing but it will be returned a few weeks later, while 3.5 million livres in cash will be left untouched. This does not mean you will leave empty-handed; 700 rifles and muskets, plus two cannons, are seized and carried outside, where the increasingly restive crowd is busy denouncing the mayor—JEAN-SYLVAIN BAILLY, an astronomer whose specialty is the moons of Jupiter—along with the Flanders Regiment and the king.

By 11am, the mood will be turning increasingly ugly, the women chanting, "A Versailles, ou à la lanterne." This is no idle threat. It has become customary to hang enemies of the people from lampposts (*lanternes*) before chopping off their heads. On hand to help is Mathieu Jouve Jourdan, better known as JOURDAN

WOMEN ON THE MARCH. TWELVE MILES IS A LONG WAY
TO DRAG A CANNON IN THE RAIN.

COUPE-TÊTE (Jourdan the Head-Cutter). Dressed all in black, carrying an array of hatchets and cleavers, Jourdan is a former butcher who gained his fearsome reputation thanks to his involvement in the murder of the financier Joseph-François de Foulon. Jourdan removed his head after his corpse had been dragged naked through the streets, then turned his attention to Foulon's hapless son-in-law, Bertier. Having shown Bertier his grisly handiwork, Jourdan proceeded to carve him into little pieces.

As the women string up ABBÉ LEFÈBVRE, the Hôtel de Ville's quartermaster, and Jourdan sharpens his blades, STANISLAS MAILLARD, a hero of the Sacking of the Bastille, intervenes and persuades them to cut Lefèbvre loose. Thanks to his standing with the crowd, Maillard is able to convince them that the time has come to march on Versailles.

THE MARCH

With Maillard and a group of drummers leading the way, you and the women begin the TWELVE-MILE MARCH in pouring rain that will persist throughout the day. You will cross the River Seine at the Cité, follow the Quai des Orfèvres to Pont Neuf, cross over again to the Louvre, pass the Tuileries Garden, halt briefly at the Place Louis XV, then continue along the Champs-Elysées and the Right Bank until you reach Chaillot, then Sèvres (about halfway), and finally VIROFLAY, the last stop before Versailles. En route you will sing POISSARD SONGS about getting your hands on *bon papa* (the king), and watch gobsmacked onlookers being corralled into joining you, while shopkeepers hurriedly board up their premises.

AT VERSAILLES

After six hours on the road you will reach the town of VERSAILLES and proceed down its main boulevard, the AVENUE DE PARIS, where you will be met by local officials who greet you with fine words and barrels of wine. You will also notice the dramatic appearance of THÉROIGNE DE MÉRICOURT, a strikingly beautiful young woman astride a jet-black horse, wearing a plumed hat and crimson riding coat, and brandishing a pistol and sabre; this former courtesan will soon become a totemic figure, embodying the spirit of women's liberation.

While most of you will mill about drinking, a detachment of women, supported by men dragging one of the cannons, will continue down the Avenue to the PLACE D'ARMES, directly in front of the Palace, only to find the iron grille gates of the COUR ROYALE (Royal Courtyard) firmly shut and protected by the Gare du Corps (Royal Bodyguard), the Cent-Suisses (Hundred Swiss), and members of the detested Flanders Regiment. Undeterred, the women will threaten to open fire, and they insult the king for refusing to sign the Declaration of the Rights of Man.

LOUIS XVI, who has rushed back from hunting at Meudon,

THÉROIGNE DE MÉRICOURT WAS THE MODEL FOR THE FIGURE OF LIBERTY IN DELACROIX'S REVOLUTIONARY PAINTING, *LIBERTY LEADING THE PEOPLE*

agrees to receive a small group of you. Once in his presence, one of the women, a seventeen-year-old flower girl, PIERRETTE CHABRY, will be so overawed that she faints; the king will rouse her with smelling salts and promise that any grain currently en route to Paris will be delivered immediately.

His offer will be met with healthy scepticism by the women outside and, as time passes, tempers fray after repeated attempts to force the gates are thwarted by the FLANDERS REGIMENT. The king, understandably worried about further enflaming the women, has the PALACE BAKERY emptied and its contents distributed among you—tuck in to their excellent bread, as this may be all you get to eat for a while. Around 6pm, the king will declare that he is ready to SIGN THE DECLARATION OF THE RIGHTS OF MAN and ratify the AUGUST DECREES that dismantled feudalism. For the moment, these concessions will mollify the crowd, and for the next few hours a sort of calm will prevail, broken only by the odd exchange of gunfire.

THE SALLE DES MENUS PLAISIRS

While you may be content to loiter outside the gates, there will be plenty to see at the SALLES DES MENUS PLAISIRS, a complex of buildings with huge rooms reserved for public functions and ceremonies, located at 22 Avenue de Paris, not far from the Palace. The Salle you are heading for was expanded to accommodate the Estates General when it opened here on May 5, 1789, but soon became home to the National Assembly. One hundred and fifty feet long, 75 feet wide, and thirty feet high, its domed roof, painted ceilings, Grecian pillars, and racked circular seating that surrounds the vast debating floor take your breath away.

At first only a few of you will accompany MAILLARD into the Salle, where he will present your demands to the bemused delegates. But then hundreds of marchers—stinking of mud and dripping wet from the long march—enter the auditorium, some with hunting knives and swords dangling menacingly from their skirts. Pandemonium ensues. Deputies are elbowed aside to make room for everyone on the benches; one woman will occupy the president's chair and issue decrees on his behalf; others will fire their muskets into the air, the shots booming round the cavernous hall. You will see a young cleric trying to stop a fishwife from further insulting an archbishop by bending and kissing her hand; she dismisses his goodwill gesture and, instead of responding in kind, tells him she will not "kiss the paws of a dog."

Not all the assembly are treated with the same contempt. The women will happily mingle with MIRABEAU, the corpulent pock-marked revolutionary whose thunderous oratory and quicksilver mind have already made him famous, while ROBESPIERRE, a thin, stiff-backed provincial lawyer who will later gain notoriety as the architect of the Terror, exchanges pleasantries and makes encouraging noises.

OUTSIDE THE PALACE: MIDNIGHT

Towards midnight you will hear the tramping of massed boots and see thousands of torches approaching through the darkness. It's the NATIONAL GUARD, nearly 20,000 of them, marching six abreast down the Avenue de Paris, wearing their red-trimmed uniforms with blue coats, white lapels, and leggings, the colors matching their cockades. This semi-professional militia, formed to defend the Revolution, is led by LAFAYETTE, a French hero of the American War of Independence and the nearest thing France has to a commander-in-chief, mounted on his white horse. Following in their train are thousands of armed civilians. The whole cavalcade left Paris at 4pm, delayed by Lafayette, who eventually gave up trying to persuade his men not to march on Versailles when a few of his officers suggested he might like the view from the top of a lamppost.

On arrival, Lafayette quickly secures an audience with the king, leaving you to wait impatiently for his return. When he finally reappears outside, he brings good news: the Flanders Regiment will be dismissed. The majority, by now a little weary, will be con-

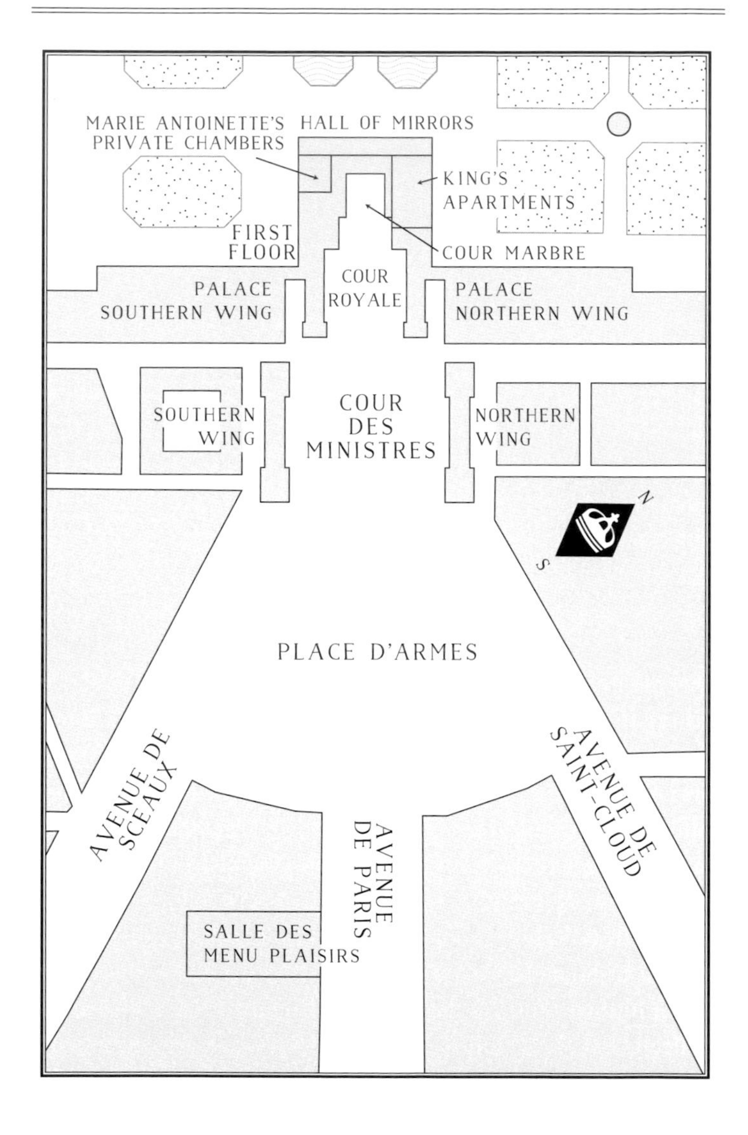

tent with this, and by 3am Lafayette, sufficiently reassured that the worst is over, will head to his grandfather's house nearby to get some shut-eye. However, almost as soon as he is gone, a seventeen-year-old apprentice cabinetmaker, JÉRÔME HÉRITIER, will be shot dead by a Black Musketeer firing from a Palace window.

BACK AT THE NATIONAL ASSEMBLY

As the night wears on, you will receive regular UPDATES FROM THE PALACE, each concession by the king met with wild applause, and be treated to a brief visit from Lafayette. However, the excitement will start to fade as exhaustion takes hold. Many women will simply crash out where they sit; some will even fall asleep standing up. You may also be ready to curl up somewhere and take a nap. But, whatever you do, don't oversleep. Make sure you are BACK AT THE PALACE BY 4AM, for things are about to kick off.

INSIDE THE PALACE

For some reason, in the middle of the night a large detachment of royal troops will be sent to the far end of the Palace Gardens, leaving the COUR DES MINISTRES lightly patrolled. The women seize their chance, breaching the gates and flooding into the COUR DE MARBRE, the Palace's innermost marbled courtyard, with direct access to the ROYAL APARTMENTS. These are located on the first floor of the immense Pavilion Block—the wings alone are half a mile wide—which also includes the private apartments of courtiers and government officials, their rooms enfilading off each other.

At this point, at around 4:40am, a few women will leave the main pack and set off HUNTING FOR MARIE ANTOINETTE; some are intent on using their aprons to carry her bowels, while one woman threatens to tear out her heart, cut off her head and fricassee her liver. Others will roam through the rooms of France's wealthiest individuals, stealing valuable items—tapestries, candlesticks, gold trinkets, porcelain figures. Look out for the DUC D'ORLÉANS, owner of the Palais Royal, leaning on the main stairway wearing a grey frock coat and slouched hat, riding whip in hand, pointing you in the direction of the royal bedrooms. As you stream towards them, you may spot terrified courtiers hiding under chairs and diving behind sofas.

Approaching the QUEEN'S CHAMBERS, you will see the way barred by two bodyguards, TARDIVET and MIOMANDRE DE SAINTE-MARIE. Tardivet makes the fatal error of opening fire and hitting somebody; he is instantly seized and decapitated. Try not to be distracted by this gory scene; instead keep your eyes on Miomandre, who is backed up against the entrance to the queen's antechambers. You will see him momentarily open the door a crack to speak to one of Marie Antoinette's maids, telling her to "Save the queen, they are here to assassinate her!" Then, seconds after he pushes the door shut again, he will be overwhelmed by the mob, who cave in his skull with their rifle butts.

A gang of women will now rush into Marie Antoinette's private rooms. Finding them empty—the Austrian Whore, wearing only her petticoat, has already slipped away and made it safely to the OEIL DE BOEUF (Bull's Eye), a locked chamber lit by a circular window, to be reunited with the king and her two children—they will vent their rage on her bed, puncturing it with their blades.

The majority of you will maraud through the magnificent HALL OF MIRRORS—seventeen arches, each with twenty-one mirrors—in search of your prey, until your progress is blocked by National Guardsmen led by LAZARE HOCHE, who form defensive ranks, some barricaded behind upturned tables, muskets at the ready. This show of force will bring you all to a standstill.

THE PALACE GARDENS

If this level of violence and mayhem is too much to stomach, we suggest you take a tour of the comparatively deserted PALACE GARDENS. Stroll past the geometrically perfect flowerbeds, cross the Allée d'Apollon, with its fountain, and you will reach the centrepiece of the park, the GRAND CANAL, a massive elongated pond in the shape of a cross, a mile long and 200 feet wide. The right arm is called the BRAS DE LA MÉNAGERIE and points towards the king's collection of wild and exotic animals; the left one is the BRAS DE TRIANON, aiming in the direction of the Grand and Petit Trianons—neoclassical chateaux—that dominate this side of the park. Marie Antoinette keeps her special commode, a state-of-the-art flushing toilet, in the Petit Trianon.

Wander in any direction through the wooded areas and you will encounter remarkable constructions: the GROTTE DE THÉTYS, a grotto dedicated to the sea nymph decorated with shells to resemble an underwater cave; a secluded grove housing the BOSQUET DES SOURCES, a circular edifice with thirty-two arches and twenty-eight fountains; the secluded SALLE DE BALLE, an amphitheater with a mini-waterfall; and the GALERIE DES ANTIQUES, a monumental array of ancient sculptures imported from Rome. However, don't wander too far afield; you'll want to be back in front of the Palace by 9:30am for the dramatic finale.

THE BALCONY SCENE

While the situation inside the Palace is at a stalemate, outside the cheering crowd has grabbed a couple of Black Musketeers and are preparing to disembowel them, with Jourdan Coupe-Tête directing operations. The HEADS OF THE BODYGUARDS Tardivet and Miomandre are, meantime, being paraded around on pikes, Miomandre's carried by an artist's model who is dressed in pseudo-Roman robes. The fun and games will be interrupted by LAFAYETTE on his white charger. Furious, he dives into the melee, frees the Black Musketeers and restores some kind of order.

Taking command, Lafayette instructs the National Guard to occupy the Place d'Armes, the Palace courtyards, and the entrance to the Avenue de Paris, then heads inside the Palace where he is admitted into the sanctuary of the Oeil de Boeuf. Here, he tries to convince the king and queen to return to Paris. Reluctant at first, they eventually bow to the inevitable. Nevertheless, it still takes Lafayette nearly two hours to get an agreement, leaving you and

LAFAYETTE PREVENTS POSSIBLE REGICIDE BY KISSING THE HAND OF MARIE ANTOINETTE.

the vast mob gathered in the Cour de Marbre anxiously awaiting the outcome, fearful that the morning will end in a blood-soaked confrontation with the National Guard.

But thankfully, at 10am, Lafayette will walk out on the balcony above you, accompanied by the king. Here LOUIS promises to come back to Paris and take up his responsibilities—and the crowd bursts into spontaneous applause. Next, Lafayette will go some way to redeeming the royal bodyguard by pinning the tricolor cockade to one of their hats; again the crowd will respond positively. Then comes the trickiest part of the performance as MARIE ANTOINETTE comes out to face the music. Hoping to win your hearts, she brings her children with her, but at the sight of them everyone will roar "Pas enfants!" ("No children!").

Earlier this year, on June 27th, Marie Antoinette had stood on the same balcony and introduced her children to a similarly huge gathering of citizens who were celebrating the birth of the National Assembly, and the crowd responded with genuine affection. But not today. An awkward, tense silence fills the courtyard. Lafayette, realising that one more false move will trigger a riot that will be impossible to stop, bows theatrically before his queen and kisses her hand. Disaster is averted, and you and the mob will once again signal your approval.

RETURN TO PARIS–AND DEPARTURE

You and 60,000 other people will leave Versailles at 1pm. At the front and rear of this snaking, unwieldy procession will be the National Guard; at its centre, the solid-gold ROYAL CARRIAGE carrying the king, the queen (who has brought a casket of diamonds with her), and their children, escorted by Lafayette; behind it are coaches filled with courtiers, ministers, and National Assembly delegates. The rest of you, a seething multitude, will ebb and flow around this parade of notables, offering protection to the train of wagons and carts filled with precious flour from the Palace stores.

The mood will be buoyant. Women are holding their swords aloft and heartily declaring that they are "bringing the baker, the baker's wife, and the baker's boy," while National Guardsmen, entering into the spirit of things, put loaves of bread on the ends of their raised pikes. Also being transported to Paris on the end of pikes are the heads of the two unfortunate bodyguards, Tardivet and Miomandre, now sporting powdered wigs supplied by a wig-maker from Sèvres; they will both end up on display at the Palais Royal. As the huge throng passes through villages on the outskirts of Paris, some locals think it best to remain behind locked doors, while others come out to watch. You will notice groups of irate peasants throwing clods of mud at the royal Carriage.

THE UNFORTUNATE TARDIVET AND MIOMANDRE RETURN ALOFT TO PARIS.

By 6pm you are back in the PLACE DE GRÈVE, by now jam-packed with people, where you will watch the royal couple enter the Hôtel de Ville, wait while they are received by Mayor Bailly and representatives from all of Paris's sixty districts, and then see them make another balcony appearance, inspiring yet more jubilation and cries of "Vive le Roi!" The roi in question, and his bedraggled wife, will leave soon after for their new home in the Tuileries.

By now an immense street party is in full swing; Paris is literally jumping for joy. Plunge into the ecstatic festivities that will last until dawn, then make your way back to your HOTEL, from where you will DEPART.

The Fall of the Berlin Wall

NOVEMBER 9–11, 1989 ✵ BERLIN

"WHAT HAPPENED IN BERLIN LAST WEEK was a combination of the Fall of the Bastille and a New Year's Eve blowout of revolution and celebration." That was the verdict of *Time* magazine in the immediate aftermath of the Fall of the Berlin Wall. In three extraordinary days the entire postwar history of Europe was rewritten and the Cold War in effect brought to an end. After nearly thirty years of longing, over a million East Germans poured through the Wall into West Berlin, turning the city into a vast street party of reunification, joy, and wonder. You really have to be there.

BRIEFING: COLD WAR

At the end of the SECOND WORLD WAR, Germany was divided into four sectors—one for each of the occupying powers (the United States, Soviet Union, Britain, and France). BERLIN, lying deep inside the Soviet sector, was divided in a similar manner. Then, as Cold War

THE WALL'S "DEATH STRIP," VIEWED FROM WEST BERLIN IN 1986. GRAFFITI ON THE WESTERN SIDE IS BY THE ARTIST THIERRY NOIR.

tensions turned to active hostility, the US, British, and French sectors combined to form the FEDERAL REPUBLIC OF GERMANY (FRG), or WEST GERMANY. Simultaneously, the Soviet sector became the GERMAN DEMOCRATIC REPUBLIC (GDR), or EAST GERMANY. The old German capital of Berlin went the same way, creating the tiny enclave of WEST BERLIN inside communist East Germany.

Over the next few years a steady stream of migrants from East Germany—above all, the young, skilled, and ambitious—headed for West Berlin and never left the FRG, thus placing an intolerable strain on the economic and political stability of the GDR. The state responded at midnight on August 13, 1961, by building the BERLIN WALL, known to the communist regime as the ANTI-FASCIST PROTECTION RAMPART. The wall, nearly 200 km long, would eventually encircle West Berlin, combining a wall on the border itself and an inner set of barriers within East Germany, fortified and patrolled with minefields, guard dogs, and a shoot-to-kill policy. Nonetheless, many would attempt to escape across the wall, and more than a hundred East German citizens would pay with their lives.

Twenty-eight years later, the Wall still stood, but the regime that maintained it was faltering, its citizens' loyalty exhausted by state tyranny, its economy a pale shadow of what was clearly visible on television beaming in from the West. The arrival of the reforming President Gorbachev in the Soviet Union had seen a wave of reforms begin across Eastern Europe. In early 1989, the opposition trade union Solidarity formed a government in Poland, while in Hungary the Communist Party initiated a massive series of reforms, opening its borders with Austria and refusing to police East German and Romanians who attempted to cross it. Over half a million East Germans would use this escape route, until the GDR responded by barring travel to Hungary. This encouraged another exodus of its citizens to the West German embassies in Prague, Warsaw, and East Berlin, where they demanded asylum. In early October, sealed trains carried thousands of East Germans to the West, and in desperation the

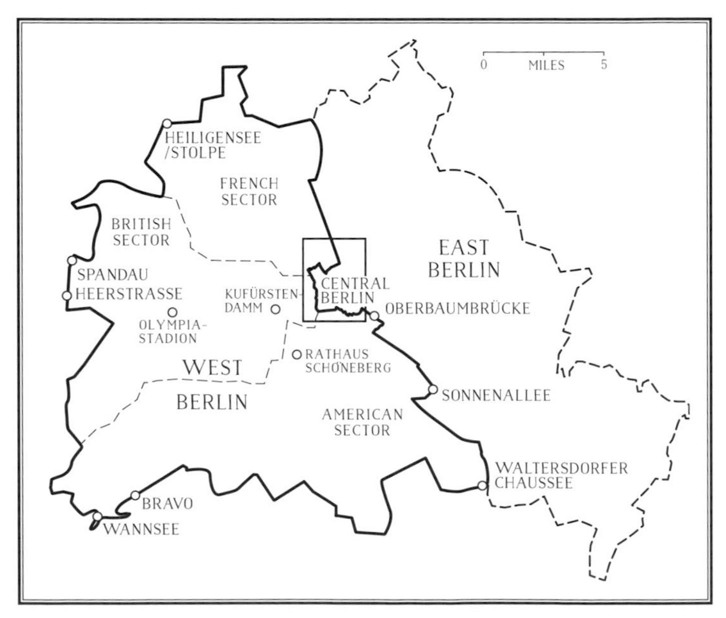

East Germans closed their borders entirely. Its citizens were left with the choice between loyalty and voice—and, over the next month, despite decades of morale-sapping surveillance and repression, they chose voice.

The protests began in Dresden, where a demonstration around the railway station turned into a full-scale riot. In Leipzig, on November 4th, the Monday Marches, organised by tiny opposition groups and the Lutheran Church, swelled into a gathering of 100,000 led by Kurt Masur, the city's orchestra conductor. The determined and sombre Leipzigers marched the entire length of the city's ring road in a display of oppositional resolve.

Two days later in East Berlin, while an increasingly panicky Communist Party met in conference to discuss regulations on movement, over half a million citizens flooded into the centre of East Berlin to oppose them. In a complex series of events guided by misunderstanding, incompetence, and panic, it was left to the hapless GDR minister GÜNTER SCHABOWSKI to give a press conference on the regime's response, live on television in both East and West. This took place at 6:53pm on November 9th. Unable to formulate the precise nature of the regime's policy, he blurted out that the new regulations (intended to be introduced very gradually) would begin immediately and be applicable to East–West Berlin crossings (also not planned in the original documents).

It no longer mattered. The people heard that the borders were open—and they acted upon the news.

THE EVENT

You will be arriving in EAST BERLIN at 6:00pm on November 9, 1989, in the hallway of 15 Finlandstrasse. The building is part of a whole street of empty houses cleared by the East German authorities after the Wall was built. You will find yourself about 200 yards from the BORNHOLMERSTRASSE CHECKPOINT: step out onto the street, turn right, and you will see the neon-lit watchtower and high wire fences of the border crossing. Please return to the hallway of 15 Finlandstrasse by midnight on the evening of November 11th. All three days of your trip will be bright, sunny, and cold, with temperatures close to zero at night. Do wrap up, and stout footwear is recommended.

ROOMS, FOOD, AND COUNTERCULTURE

A room in East Berlin has been booked for you at the HOTEL METROPOL on Friedrichstrasse, just north of the city's most famous boulevard, Unten den Linden. Opened in 1977 as the modernist flagship of *Interhotel* (the East German state tourist agency), the hotel is reserved exclusively for Western visitors and accepts only hard foreign currency. Its 1970s decor—orange plastic fascias and brown-patterned wallpapers—are a bit tired, but it remains a pleasant and tranquil oasis for this tumultuous weekend. Do note that all rooms are extensively bugged, so be discreet in any reference to the agency, or indeed time travel or events as-yet-unfolded. For the more adventurous, a roof over your head is also available amongst the SQUATS of PRENZLAUER BERG, heartland of East German counterculture; try the crowd at 61 Lychenerstrasse or 7 Beherferlinerstrasse.

While in East Berlin, we strongly advise travelers to sample local food and drink. Nothing will taste the same again in post-Wall Germany. The hot dogs, in particular, have a sinister color and consistency; the fizzy drinks are weird chemical conconctions; even the schnapps ain't right.

We expect most visitors to focus on the action near the Wall, but if you want to take in a little of WEST BERLIN'S COUNTERCULTURE, we suggest you make for the PIKE CLUB, in a back courtyard on Heinrich-Heine Strasse, in the Kreuzberg quarter, just south of Checkpoint Charlie. On the night of the 9th, East German punk outfit DIE ANDEREN (The Others) are playing; oddly, a number of East German bands—even punks—have had permits recently. By the time you arrive it will be a beer-soaked, pogoing, manic punk party. Also well worth a visit over the weekend is DSCHUNGEL, a modernist 1920s café that pulls in the most louche cocktail drinkers in the city; recent guests include David Bowie and Iggy Pop.

THURSDAY, NOVEMBER 9TH: DIVIDED BERLIN

Having located the BORNHOLMER STRASSE CHECKPOINT, take a little time to wander the streets of the MITTE quarter with its fabulous crumbling architecture. Since the building of the Wall, the area has steadily emptied of its original working-class population and become a magnet for the most disaffected of the

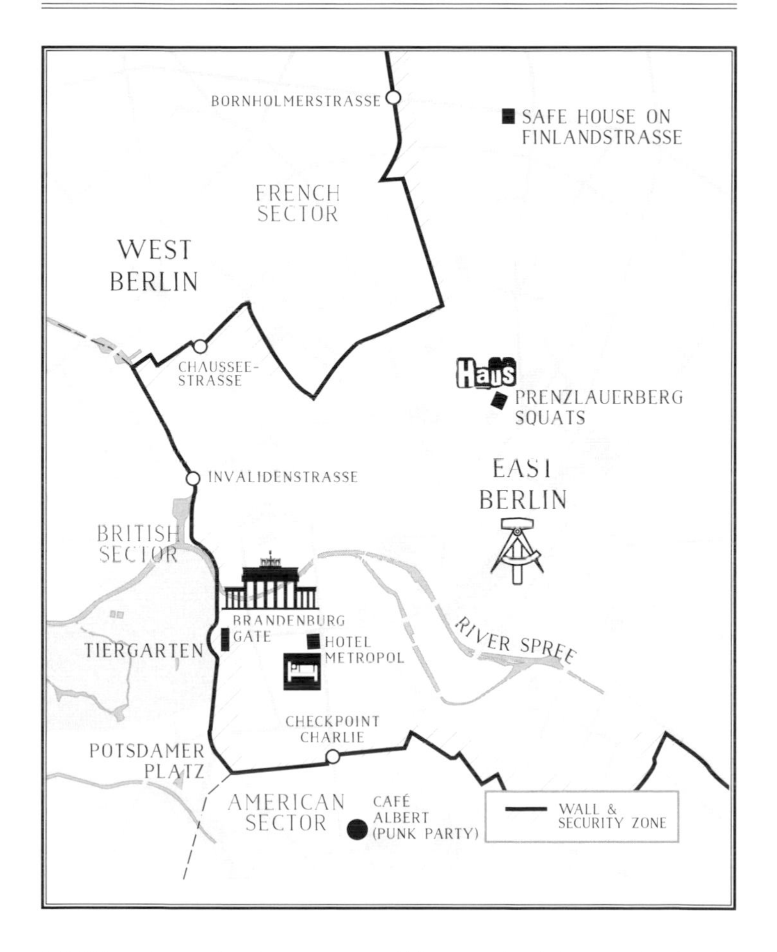

GDR's citizens. In the many small bars and drinking holes in the streets around the checkpoint you will hear the locals discussing the meaning of the Schabowski press conference and wondering whether the border really might be opened this evening.

Do not linger too long. From around 7:20pm onwards you will find a crowd beginning to gather at the Eastern entrance to the Bornholmer Strasse checkpoint. Two particularly vociferous young men—ARAM RADOMSKI and SIGGI SCHEFKE—will be shouting out questions to the perplexed-looking border guards, asking whether they can cross over to the West. The guards will be joined by a senior officer, more discussion will ensue, but no permissions to cross will be given at this point. If you can, try to look up into the brightly lit offices of the checkpoint where the figure of LIEUTENANT COLONEL HARALD JÄGER, the senior officer on duty this evening, will occasionally appear. He is square-faced, with a side parting, and, as you wait at the gate, he will be furiously phoning his superiors for advice on what to do as the crowd grows in size and in its sense of anticipation.

By 8:00pm the crowd will have swelled to a few hundred and will be joined by an increasingly long line of vehicles full of people seeking to cross. At 8:30pm a police car will work its way to the front of the crowd. The officer in that car will address the crowd by megaphone, instructing them to go to a nearby police station to gain an exit visa. You should ignore this instruction, as will most of the locals. At 9:00pm the border guards will reappear and take the loudest and most persistent members of the crowd into the CHECKPOINT OFFICE. Following orders from an increasingly befuddled Stasi HQ, the plan at this point is to stamp their passports, show them the way to the West, and expel them permanently from the country. Around thirty or so people will be let through on this basis; you should not try and join them. Stay with the crowd, now many thousand strong, reaching back down Bornholmer Strasse and spilling into the side streets. Listen out for the regular chants of "Open the gates!"

At approximately 11:30pm, Harald Jäger, on his own initiative, will give the order to OPEN THE GATES. Two of his subordinates, Helmut Stoss and Lutz Wasnick, will appear by the main barriers and begin pulling them open by hand. Immediately the huge crowd behind the gate will start pushing forwards and force the gates open themselves. A great surge of humanity will follow as thousands of laughing, crying, shouting, and jubilant East Berliners head through the gate and into the West. As you go, it is worth pausing to catch the looks of weary incomprehension on the faces of the border guards.

ROOM FOR ONE MORE. THE BRANDENBURG GATE, NOVEMBER 9TH. THE TEXT ON THE SIGN, "NOTICE! YOU ARE NOW LEAVING WEST BERLIN," HAS BEEN MODIFIED WITH THE WORDS "WIE DENN?" ("HOW, THEN?').

INTO THE WEST

Once you have passed through the Bornholmer Strasse checkpoint, cross the BÖSEBRÜCKE BRIDGE, and you have arrived in West Berlin. You can stop here and watch the extraordinary lava flow of East Germans coming through all night. Some West Berliners will be gathering to watch and to greet their fellow countrymen, but we suggest at some point in the evening you take a stroll to see the action at other points along the Wall.

Approximately fifteen minutes' walk southwest is the CHECKPOINT ON INVALIDENSTRASSE. You will find that the Stasi and the East German border guards here are taking a much stiffer line with East Germans trying to cross over, as indicated by the large, noisy, and agitated crowd that are stuck on the other side of the border. West Berliners will also be gathering here. Around 1:00am, look out for the arrival of WALTER MOMPER, the Social Democratic mayor of West Berlin. He is easily recognisable—a bald man with a very bright-red scarf and a megaphone—and will be addressing both the crowd and the West and East German border guards, calling for calm and for the East Berliners to be let through. By 1:30am it is pretty clear to all that no one will be passing across here this evening, so take the first left off Invaliden Strasse and head into the TIERGARTEN, West Berlin's large park. You can't miss UNTER DEN LINDEN, the large road that runs east–west through the park. Follow the crowds east, and after 400 yards the Wall should come back into view. Behind it you will see the great neoclassical columns of the BRANDENBURG GATE.

Although there is no checkpoint at the Brandenburg Gate, which is inaccessible from both East and West Berlin, lying in no-man's land between the two parts of the Berlin Wall, Berliners from both sides of the divide have been drawn to the gate all evening. The scene will be all the more dramatic as American television companies have established themselves here and installed floodlights to illuminate both the wall and their own broadcasts. On the Western side the Wall is unusually low and has, for some of its length, a flat top. If you arrive early you will find a crowd just milling around, but from 9:00pm onwards individuals will begin climbing up to the Wall. We advise that you not join them at this point, as the East German border guards will be using water hoses for the next couple of hours in an effort to knock people off. Look out, though, for one young man who will be braving the water cannon, shielding himself using an umbrella given to him by one of the crowd. Later in the evening the hoses will be turned off and the Western side of the Wall will make an excellent vantage point from which to view a steady stream of East Berliners walking across the eerily lit plaza that the Brandenburg Gate stands in.

Your final stop on the evening's itinerary is CHECKPOINT CHARLIE, the most famous of the border crossings, located in the US zone of West Berlin, south and east of the Brandenburg Gate. Just follow the line of the Wall until you reach Friedrichstrasse. Here you will see the small metal hut that serves as the checkpoint office on the Western side and the famous CAFÉ ADLER, a regular haunt of spies and military officials during the Cold War. Should you choose to come early in the evening, café owner Albrecht Rau can be seen making his way to the East with a tray laden with glasses, sparkling wine, and hot coffee for the border guards. They will refuse the drinks, but Albrecht will be celebrating anyway. The border guards will then attempt to seal the border by putting a variety of additional barriers in place, but by a quarter past midnight the crowds will start tumbling over them.

FRIDAY, NOVEMBER 10TH: CENTRAL BERLIN

Today is a day to just wander around central Berlin. Over half a million people will be making the journey through the Wall today. The vast majority will do so on foot, but some 26,000 vehicles will cross the border, nearly all of them East Germany's characteristic TRABANTS and WARTBURGS. You will soon become accustomed to the unique sound of the Trabant's sclerotic two-stroke engine and the peculiar fumes and billowing toxic smoke generated by the ersatz mixture of petrol and oil that they run on. Some cars will be flying the East German flag but with the hammer-and-sickle emblem at the centre cut out, leaving it remarkably similar to the West German flag.

You will also note the sharp differences in attire between East and West Berliners. VISITORS FROM THE EAST have unsurprisingly drab wardrobes and thin coats, economic inequalities that will persist for the next three decades. This will all be nicely summed up in POTSDAMER PLATZ mid-morning when the West German president, RICHARD VON WEIZSÄCKER, will arrive, shake a few hands, and then head off in his gleaming black Mercedes followed by a dirty-khaki Wartburg and a filthy-green Trabant. Meantime, take in the sweet sounds of a hundred HAMMERS AND CHISELS at work taking souvenirs of the Wall—in the West, the most covetable areas are those covered in graffiti. This is also a moment to appreciate the speed at which the commercial impulse can work: much of the new chippings and rubble will soon be turned into souvenirs, while t-shirts with the legend *Ich war dabei: November 9* (I was there: November 9) will already be on sale near the Wall.

But today is about unity. Everyone is a Berliner and everyone is a German, including the enormous number of people flooding into the city from all over Western Europe. One example not to miss is the Amsterdamer in Postdamer Platz who will be parking up with a van full of 10,000 roses straight from the Dutch city's famous flower market. He will proceed to give them to the new arrivals. Flowers are not the only gifts on offer. Public transport is effectively free today and tomorrow, with bus drivers refusing fares and the underground system full to bursting. A special edition of the local newspaper, *Berliner Morgenpost*, will be given away free on the streets, and throughout the day many individual Berliners, and some cafés and restaurants, will be offering free drinks, snacks, teas, and coffees. Near the checkpoints there are trucks doling out bags of free fruit, chewing gum, and Western cigarettes—another treat that will be spontaneously passed around. For those that are partial, the Kreuzberg district around Checkpoint Charlie also offers plenty of welcoming joints. Follow your nose and look interested.

On their first visit to the West, East Germans are entitled to a welcoming gift from the West German government of 100 deutschmarks, available on presentation of ID at any bank or post office. The BERLINER BANK on West Berlin's fanciest shopping street KURFÜRSTENDAMM is worth looking in on to see this in action—expect a queue five deep and a quarter of a mile long. While the street's enormous and opulent department stores will be drawing astonished looks from East Berliners, they won't be spending much of their money there. Many, however, can be seen holding bags of fruit—bananas, oranges, and kiwis—hitherto unavailable to them.

There are two special events today worth making a detour for. For the political junkies, West Germany's leading politicians will be speaking at RATHAUS SCHÖNEBERG at about 2:30pm. Originally merely the town hall of the Schöneberg district, the building has been the seat of the Berlin Senate and the mayor's office since 1950—Berlin's traditional seat of government, the Rotes Rathaus lies in ruins still in Mitte on the other side of the wall. The cast includes West German chancellor HELMUT KOHL, only recently arrived in the city after cancelling his official visit to Poland, and his foreign minister HANS-DIETRICH GENSCHER; WILLY BRANDT, a native of the city and chancellor of West Germany in the 1960s and early 1970s; and WALTER MOMPER, the current mayor of West Berlin. It will be an old-fashioned media scrum as the photographers, cameramen, and boom operators jostle at the front. Kohl's key moments come when he says, "Long Live a free German fatherland! Long live a free Europe," and "I want to call to all in the GDR we're on your side, we remain one nation. We belong together." Willy Brandt, visibly moved, will offer a moment's caution: "This is a beautiful day after a long voyage, but we are only at a waystation. We are not at the end of our way."

To get to the Rathaus Schöneberg, take a train on Line 2 heading west from the centre and change at Nollendorf Platz for Line 4; you'll be looking for a train heading south to Innsbrucker Platz.

Music lovers should head for the section of wall just to the West of CHECKPOINT CHARLIE. There you will find the great Russian cellist MSTISLAV ROSTROPOVICH playing. Rostropovich defected from the Soviet Union in 1971 and has just flown in from Paris to offer this impromptu concert in honour of those who died trying to cross the Wall. He will be opening with a sensational rendition of Bach's second suite for cello, the "Sarabande."

SATURDAY, NOVEMBER 11TH: A GAME OF FOOTBALL

Saturday continues very much in the vein of Friday, with ever-larger numbers of East Germans flooding across the now routinely open border crossings, still being met by ecstatic West Berliners. New crossings will be opening today at the GLIENICKE BRIDGE and on EBERSWALDER STRASSE, where an East German army bulldozer will actually smash down a segment of the Wall to relieve pressure on the other checkpoints.

Mid-morning will see a more DIY effort to do the same in Potsdamer Platz, where a group of ALT-WEST BERLINERS will be showing up in an old jeep. Parking up close to the Wall, they will attempt to fix a huge chain to the concrete slabs and then to the jeep. Around them the crowd will take up the cry, "The Wall must go!" as well the now ubiquitous "One land, one people." The bohemian feel of the moment is enhanced by a group of pan-pipe players nearby. These efforts will be thwarted, however, by East German border guards turning a water cannon on the jeep crew. They will be ably supported by the West German police, who will park a phalanx of vans up against the Wall to prevent a reoccurrence of the event. The crowd will respond to the East German guards, now on the Wall itself, with cries of "Come down! Do you want to starve over there?"

ROSTROPOVICH PLAYS BACH IN FRONT OF THE WALL: NOTE THE GRAFFITO "CHARLIE'S RETIRED." THEY DON'T WAIT AROUND WITH THE SPRAYPAINT.

Alternatively, head for the FOOTBALL MATCH between HERTHA BERLIN and WATTENSCHEID kicking off at 3:00pm at the Olimpiastadion, the centrepiece of the 1936 Berlin Olympics, in the west of the city. Prior to the opening of the Wall, it was billed merely as a top-of-the-table clash in the German Bundesliga 2, but since the morning of November 10th it has become a remarkable opportunity for all of Berlin to celebrate the historic moment.

Until 1971 Hertha Berlin were located in the Wedding district of the city, at their old stadium, the Gesundbrunnen. In 1961 the Wall was erected just a few hundred yards east of the stadium, closing it off to the large body of support in East Berlin. For many years, East Berliners would gather near the Wall to hear the sounds of the crowd as Hertha played on the other side. Today they have the chance to go and see them in their new home, the OLIMPIASTADION, with 10,000 free tickets available to anyone who can produce an East German ID card. Don't worry if you can't, though; there will be plenty of opportunity to just show up and buy a ticket.

Prior to the kick-off, the stadium announcer will call out the names of all the districts of Berlin—both East and West—and the crowd will respond heartily. Wattenscheid will take the lead after sixteen minutes of the first half, but in the sixty-fourth minute, nineteen-year-old Sven Kretschmer will grab a scrappy equaliser for Hertha. The team will go on to win promotion to the Bundesliga's first division at the end of the season.

To get to the game, you'll need to head south of the Brandenburg Gate to one of the stations on the U1 metro line—either Möckernbrücke or Gleisdreieck—and head west on the trains bound for Ruhleben. Olimpiastadion is the penultimate stop.

PART THREE

Cultural & Sporting Spectaculars

The 235th Olympiad

AUGUST, AD 161 ✵ OLYMPIA, GREECE

IN THE FIFTH CENTURY BC, THE POET Pindar called the Olympics the "pinnacle of contests"—and the ancient games held at Olympia, almost a millennium old when you arrive, still fit the bill. Our trip offers the chance to spend a week in this beautiful complex of temples, shrines, stadiums, and bathhouses on the rolling plains and meadows of Elis in the northwest Peloponnese. With its Olympic site recently and extensively refurbished by the Roman Empire, the 235TH OLYMPIAD offers all the ritual and the spectacle of the Classical Hellenic era, and a level of comfort and quality of accommodation that is unmatched in the anceint world. With unrivalled access to the whole sanctuary, you will enjoy a week of processions and sacrifices, pentathlon and pankration, chariot racing and wrestling; and the unique opportunity to see a real wonder of the ancient world. Note that events at the games can only be witnessed by MEN, so all travelers will be dressed and made up accordingly.

BRIEFING: THE ANCIENT OLYMPIC GAMES

The SANCTUARY OF OLYMPIA, set in the foothills of Mount Kronos in the eastern Peloponnese, has been a place of spiritual pilgrimage and worship for over a thousand years before your arrival in AD 161. As early as the eighth century, permanent shrines were built here to worship the ancient Hellenic mother gods. At some point, conventionally dated as 776 BC, RITUAL GAMES began to be played at Olympia, though they were just local affairs. In the sixth century BC these games began to attract a wider audience and take on a Panhellenic quality, while Olympia itself acquired the magnificent TEMPLE OF ZEUS, whose giant sculpture of the god was considered one of the Seven Wonders of the Ancient World. Reflecting the fanatical body culture of the gymnasiums and the barracks of Greece's growing city-states, the Games took on ever-greater significance and popularity. In the early fifth century BC, the programme of events at the Games was fixed as a five-day festival of athletics, and it remained substantially unchanged until the Romans conquered Greece in the second century and added an extra day to the party.

As with so much else about Greece, the ROMANS left things pretty much intact but improved the infrastructure. At these Olympics, visitors have truly never had it so good. Back in the day, visitors spending a week at Olympia would

SPOT THE DIFFERENCE! AN ARTIST'S IMPRESSION OF THE WALLED ALTIS AT ANCIENT OLYMPIA, SHOWING THE TEMPLE OF ZEUS IN THE FOREGROUND. YOU WILL STRUGGLE TO FIND THE LARGE AMPHITHEATER ON THE LEFT, WHICH IS ENTIRELY FANCIFUL.

find there were no public toilets and no supplies of running fresh water; they would stay in tent villages that emerged in the fields around the sanctuary, beset by the intense heat and dust of a Greek August.

Epictetus the Stoic, true to form, thought even that worth enduring: "Are you not scorched with heat? Are you not cramped for room? Is not washing difficult? Are you not wet through when it is wet? Do you not get your fill of noise and clamour and other annoyances? Yet I fancy that when you set against all these hardships the magnificence of the spectacle you bear them and put up with them." Aelian, by contrast, decreed it a fate worse than forced labour: "A man from Chios got angry with his house-slave and said, 'I shan't put you on the treadmill, but I will take you to Olympia.' For the master thought it a much harsher punishment that he should watch at Olympia, and be baked under the sun, than that he should be sent to grind corn on the treadmill." Either way, the informal Olympic Village remained riotous enough to need its own dedicated police force, known as the "whipbearers."

THE TRIP

We plan to offer future trips, for keen Hellenists, to the more ascetic pre-Roman Olympics. However, visitors on this trip will witness the games—the 235TH OLYMPIAD—in Roman style, and in comparative luxury. The Romans have already added luxurious accommodation and proper bathhouses and toilets, and this year there will be running drinking water for the first time. The site as a whole is looking fresh, too. Over the past ten years, HERODES ATTICUS, the richest man in Athens, has been building the NYMPHAEUM—a great memorial water feature dedicated to his late wife—and a plumbing system to supply it. At this Olympiad it will be in working order for the first time, making your stay infinitely more pleasant and the environment infinitely more healthy. You will ARRIVE BY BOAT on the afternoon before the game, landing on the north bank of the KLADEOS RIVER, where you will be taken to your room in the nearby ROMAN HOTEL COMPLEX. You will be DEPARTING from here on the evening of the sixth and final day of the festival.

THE FESTIVAL SITE

At the centre of Olympia is the sacred walled compound, THE ALTIS, that contains the site's temples and shrines. The Romans have built up and repaired the ALTIS WALL, added porticos and monumental arches, and laid out paths and gardens. Normally, this would be a very tranquil place, but over the next week it is going to make the Roman Forum look quiet. Thousands of visitors from all across the Roman Empire will mingle with athletes, trainers, priests, dignitaries, poets, judges, and umpires. There will also be plenty of servants, cooks, and factotums from amongst the entourages of the elites, as well as itinerant soothsayers, conmen, and hawkers spilling over from the TENT CITIES in the surrounding meadows.

You will enter the Altis through the PROPYLON, a monumental arch set into the wall, at the northwest corner of the compound. To your immediate right will be the PRYTANEION, the nerve centre of Olympia and its games; nearby is a small circular monument, the PHILIPPEION. Built in the sixth century BC, the Prytaneion houses the high priests of Olympia, and an eternal sacred fire still burns here. During the Games, officials will use the building and entertain Olympic champions in its dining chambers. The Philippeion is a small, circular memorial, its carved marble roof held up by an Ionic colonnade. It was commissioned by Philip of Macedon to commemorate his victory at the Battle of Chaeronea in 338 BC and serves as a reminder of his victory in the horse racing at Olympia in 356 BC. Inside you will find a family scene in ivory and gold: Philip, his father; his son Alexander the Great; and his fourth wife and Alexander's mother, Olympias.

THE TEMPLE OF HERA

Standing with your back to the Philippeion, you will be able to see the TEMPLE OF HERA to your left and the PELOPION to your right. Beyond the temple, look for the tiny shrine known as the

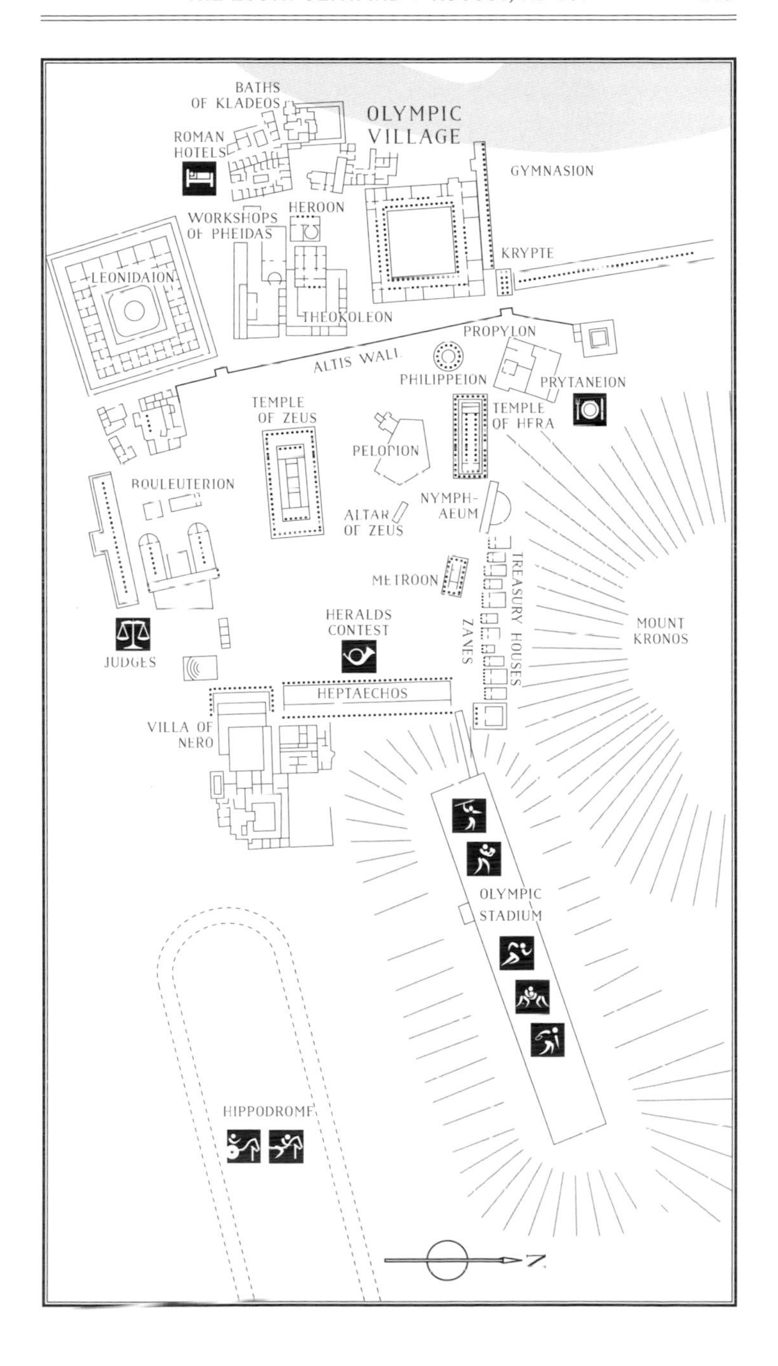
BATHS
OF KLADEOS
OLYMPIC
VILLAGE
ROMAN
HOTELS
GYMNASION
HEROON
WORKSHOPS
OF PHEIDAS
KRYPTE
LEONIDAION
THEOKOLEON
PROPYLON
ALTIS WALL
PHILIPPEION
PRYTANEION
TEMPLE
OF ZEUS
TEMPLE
OF HERA
PELOPION
BOULEUTERION
NYMPH-
AEUM
ALTAR
OF ZEUS
METROON
TREASURY HOUSES
ZANES
MOUNT
KRONOS
HERALDS
CONTEST
JUDGES
HEPTAECHOS
VILLA OF
NERO
OLYMPIC
STADIUM
HIPPODROME
N

METROON and behind it, on a raised terrace, a string of small TREASURY HOUSES. The Pelopion is already an ancient graveyard built atop ancient burial sites, now enclosed by a irregular pentagonal wall with a Doric portico for an entrance. Inside you can enjoy the tranquillity of its poplars, olive trees and statues, and its mounds of funereal and sacrificial ash. During your stay you may well see both formal and informal rites in here, especially the roasting of meat. Do be aware of the consequences of partaking. The writer PAUSANIAS, on a contemporary visit to the games, noted that "the magistrates sacrificed a black ram to honour Pelops and whoever ate from the sacrificed animal was not allowed to enter the temple of Zeus." If you have a taste for the black ram, perhaps it would be best to try to see the Temple of Zeus first.

Built in the early sixth century, the Doric TEMPLE OF HERA will appear strangely remodelled, as Roman stone and sculpture have recently replaced the old Hellenic work, including its wooden columns. During the Games this temple's main function will be to display the olive wreaths for Olympic victors on a long wooden table. Note, just to the right of the temple as you look north, a large conical mound of compressed ashes; this is the ALTAR OF ZEUS, where many centuries of sacrificial remains have accumulated to form a small hill over twenty feet high.

Immediately beyond the Temple of Hera, and most likely with an appreciative crowd around it, is the newly opened NYMPHAEUM. Built by Herodes Atticus in about 160, in memory of his late wife Regilla, a priestess of Demeter here at Olympia, it is in effect a vast monumental spring and water feature. You will see a huge oblong pool backed by a two-storey marble apse, festooned with carved figurines and topped by half-cupolas. The niches within the apse hold Atticus's ancestors and the obligatory sculptures of Zeus, while the centrepiece of the building is a huge marble bull set in the pool. However, the real genius of the Nymphaeum is hidden, for it is just the decorative frontage of a huge piece of Roman civil engineering that has, for the first time, brought fresh water from a spring a few miles east along a stone viaduct to Olympia. An equally remarkable system of storage tanks and pipes makes the water available across the sanctuary for drinking and bathing.

The nearby METROON is a small Doric temple, built in the early fourth century BC and dedicated originally to RHEA, daughter of the original Greek earth mother, Gaia. At the back, there is still a small altar to her, but over the past century or so the Metroon has moved into Roman Emperor–worship; inside the colonnade are statues of the emperors Augustus, Claudius, Titus, and Vespasian, Nero's mother Agrippina, and Domitian's wife Dominitia.

The path that leads from the Metroon to the KRYPTE, the large monumental arch in the northeast corner of the Altis, and thence to the Olympic stadium, is lined with an eclectic collection of bronze sculptures of Zeus. These are the ZANES, paid for by fines levied on cheats and transgressors of Olympian law and inscribed with warnings and admonitions for athletes on their way to the Games. The first dates from the 98th Olympiad, held in 388 BC, when the boxer Eupolos of Thessaly was fined for bribing three of his opponents. Half a century later, six were erected by the Athenian Kallipos. He had been bribing other athletes in the Pankration. By the second century AD the TREASURY HOUSES which sit on the terrace behind the Zanes are a real curiosity. Built over half a millennium ago by the small city-states of the Hellenic Mediterranean diaspora, they once housed the votive offerings of their athletes and champions.

EAST SIDE: HOUSE OF NERO

A COLONNADE occupies almost the whole of the east side of Altis. Built in the fourth century by Philip of Macedonia to celebrate his victories in the Olympic chariot races, its unusual acoustics will give rise to its contemporary Greek name of HEPTAECHOS (seven echoes). This is the place to come and hear the competitions amongst the trumpeters and heralds that are an integral part of the Olympic Games. Both were, and continue to be, part of the sporting ceremonies, announcing the beginning of a competition with a fanfare on their enormous elongated instruments, or declaiming the name, the father, and the home town of Olympic victors. In addition to these duties, they will be sparring among themselves in contests throughout the Games.

At the southeastern corner of the Altis, just beyond the colonnade, sits the VILLA OF NERO. The volatile emperor came to Olympia as part of a long tour of his Hellenic realm, a schedule that required, for the first time in centuries, that the Games be delayed for two years to accommodate him. You will no doubt hear plenty of Nero stories while you are here, notably how back in AD 67 he crashed his ten-horse chariot, and how to everyone's amazement he won all of the musical and literary competitions in which he participated. He also had an ancient Greek shrine to Hestia ripped down and put up his own brash Imperial condo. It has been enlarged since then and equipped with its own baths to accommodate the very pinnacle of the Roman elite visitors.

THE TEMPLE OF ZEUS

The largest building in the Altis and the spiritual centre of the Games is the TEMPLE OF ZEUS. The temple, started in the early fifth century BC, is considered an exemplar of Doric temple architecture. However, the building was eclipsed around 430 BC, when it received the great seated STATUE OF ZEUS carved by PHEIDAS and designated one of the Seven Wonders of the Ancient World. Before you head inside, do take a moment look at the sculptures on the triangular pediments at the east and west end of the building: the former depicts the mythical CHARIOT RACE BETWEEN PELOPS AND OINOMAOS, presided over by Zeus; the latter, the BATTLE OF THE LAPITHS AND CENTAURS, arranged around the god Apollo. Note too the lion-head water spouts on the marble roof, the gold cauldrons that sit at each corner, and the twelve panels, or *metopes*, that illustrate the twelve labours of Hercules.

THE STATUE OF ZEUS LOOKS ON APPROVINGLY AS WRESTLERS FROLIC.

Once inside you will cross a small rectangular space paved with hexagonal marble slabs: this is where the Olympic champions will receive their olive wreaths. ZEUS, of course, is at the very heart of the temple. Over forty feet high, made of gold and ivory, he won't be hard to find, but just in case: he is the guy on the throne, crowned by a wreath of olive leaves, holding a sceptre topped by an eagle in his left hand and a winged NIKE, goddess of victory, in his right. The Nike figure alone is six feet high. Enjoy it while you can, as it will be taken off to Constantinople in the fifth century and destroyed in a fire.

South of the Temple of Zeus you will find the SACRED OLIVE GROVES that provide wreaths for Olympic champions and the BOULEUTERION, or council house, which combines a central open-air hall and two wings of offices and storage rooms. There will be plenty of action here during the Games, as the building serves as the seat of the ELEAN SENATE, who are the organising committee of the Games, and the HELLANODIKAI, who are the umpires and referees. This is where athletes will be registered for competition and, when necessary, lots will be drawn to determine either their opponents or starting positions in races. The central hall will be used during your visit for hearing any accusations of bribery, corruption, cheating, or other breaches of Olympic convention. The place is dominated by a mean-faced STATUE OF ZEUS clutching some equally mean thunderbolts. Do take a look at the inscription of the pedestal, which curses perjurers and cheats.

THE OLYMPIC VILLAGE

The large collection of buildings that lie on the western side of Altis, on the plain that runs down to the river, is the OLYMPIC VILLAGE. At the northern end sits a cluster of sporting and training facilities: the GYMNASION, the XYSTO, and the PALAESTRA. Next comes a block of older buildings and shrines that currently houses Olympia's priestly caste and its workforce. Finally, at the southern end, are the new ROMAN HOTELS and the fabulous BATHS OF KLADEOS for well-heeled visitors and the connected, and their ancient predecessor, the LEONIDAION.

The GYMNASION, entered through a vast portico, is a very large quadrangle building whose central court is 200 yards long. A roofed colonnade—the XYSTOS—divides this into two, with one space for DISCUS AND JAVELIN PRACTICE, the other for RUNNING. Look for the small green door at the southern, colonnaded end of the courts; it will open out on to the PALAESTRA, a large square building set around a courtyard covered with sand. You will find it thickly populated with BOXERS, WRESTLERS, COACHES, and MEDICINE MEN. On the south side of the courtyard, the internal colonnade leads to the APODYTERION, or undressing room, while on the north side you will find the EPHEBION, or club room, for post-training relaxation and socialising. A wander around should also reveal the ELAIOTHESION, or oil store, the KONISTERION, or dusting room, and a few SPHAIRISTERIA, which are rooms for ball play. Note the strip of concrete pavement in front of the club room and its alternate pattern of ribbed and smooth tiles—this is used as a form of BOWLING ALLEY.

The central cluster of buildings in this section of Olympia is dominated by the THEOKOLEON. The *theokoloi* are the full-time priests of Olympia, attending to all its spiritual business between Olympiads—as well as presiding over many of the week's celebrations. They share their office space with a small interpreting service, full-time soothsayers, and the quarters of the sanctuary's musicians, sacrificial animal experts, and the woodsman charged with keeping the many temple fires and votive flames stocked with kindling. Despite all the powerful mystic knowledge gathered here, no one really quite knows what the neighboring HEROON is for any more. Once a small baths, then a Greek temple, it is now a cultic Roman shrine, with an altar to Pan. Finally, there are the buildings referred to as PHEIDAS'S WORKSHOP and still a centre for masonry and stonework in Olympia. Pheidas was the fifth-century Greek sculptor responsible for some of the most important statuary of classical Hellenism—above all, the bronze goddess Athena at the Acropolis in Athens and the seated Zeus that you will have seen in the Temple of Zeus here in Olympia.

The Greeks did build a few small dormitories and hostels in the third and fourth centuries, and even some rudimentary baths, but all of this has been swept away by the Romans. At the same time as remodelling the Altis and generally sprucing the place up, the Romans replaced the primitive Greek hostels with a series of grand HOTELS, installed fine mosaic floors, and built the BATHS OF KLADEOS on the banks of the river. This building, less than eighty years old by the time of your arrival, is in excellent shape, featuring grand vaulted ceilings, multicolored marble cladding, a sweat room, bathtubs, and public lavatories, not to mention Olympic-size hot and cold pools.

By contrast, the LEONIDAION, although extensively renovated by the Romans, will be showing its age—almost half a millennium by the time of your stay. It was commissioned and designed by Leonidas of Naxos in 332 BC, and his statue can be seen at the northeast corner of the building. Again it is a quadrangular structure with a central courtyard and has previously served as an athletes' hostel, but it has been commandeered by Roman officials and given a nouveau riche twist with a huge ornamental pool in the old court.

THE SPORTS FIELDS

Follow the path that runs down from the triumphal arch and you will see the OLYMPIC STADIUM in front of you. First laid out in the sixth century, this late-Roman version is about fifty yards south of the original, nestled at the foot of Mount Kronos. The arena is an oblong of sand more than 200 yards long, forty yards wide, and surrounded on all sides by grass-covered embankments large enough to accommodate 45,000 spectators. When the Games are on, most of them will be standing, though the Romans, keen as ever on home improvements, have added banks of wooden benches on part of the long sides and improved the drainage by installing a new stone water channel around the track.

In the centre of the south embankment there is a stone platform reserved for the Hellanodikai. On the north bank, note the marble ALTAR OF DEMETER, whose priestess will be the only woman permitted to watch the Games. Behind the west end of the stadium you will be able to see the APODYTERION—the undressing room—where athletes will prepare during the Games. This is a secure area.

Immediately to the south of the stadium, and three times its size, is the HIPPODROME, with a gravel track more than 800 yards long and eighty yards wide, surrounded by low artificial banks. The track itself is divided down the middle by a stone barrier, the EMBOLON, while a column at each end marks the starting and turning point of each circuit. You will see a sculpture of the Hippodamia (wife of the cheating charioteer Pelops) and a circular altar dedicated to Taraxippos, the generic ghost and horse-worrier that haunts hippodromes in the ancient world. Recent reports from Pausanias suggest that this might be an interesting place to locate yourself during the races: "There stands, at the passage through the bank, Taraxippos, the terror of the horses. It is in the shape of a round altar and there the horses are seized by a strong and sudden fear for no apparent reason, and from the fear comes a disturbance. The chariots generally crash and the charioteers are injured. Therefore the drivers offer sacrifices and pray to Taraxippos to be propitious to them."

THE GAMES

The programme of the ancient Olympics was first fixed in 486 BC as a five-day event. The arrival of the Romans has seen it extended to SIX DAYS. Please note that all of the ATHLETES will be competing NAKED, a tradition supposedly established in 720 BC by Orsippus of Megara, who won his race after his loincloth fell off and attributed his speed to this. A Spartan sprinter called Acanthus went on to win the second race naked too, sealing the worth of the innovation. Many athletes will tie tight leather thongs strings around their foreskins and thighs to prevent unwanted motion and erection.

VICTORY CEREMONIES will occur relatively soon after each event. First a trumpeter will blow a fanfare, then a herald will announce the champion as "the best in all Greece," who will then be awarded a ribbon. Informal parties, feasts, and the award of palm leaves will then break out. In the stadium, victors will also take a *periageirmos*—a victory lap. The more formal ceremonies and the presentation of the olive wreath will come later in the week.

DAY ONE: OATH TAKING

The main event on the opening day will be the gathering of the HELLANODIKAI—literally the "judges of the Greeks"—notable for their purple cloaks. As well as being in overall charge of the

Games, they are also responsible for overseeing the taking of the Olympic oath at the TEMPLE OF ZEUS. Pausanias reports that "competitors, their relatives, and their trainers swear that they would be guilty of no foul play in the Games, and judges swore that they would be fair and would not accept bribes." Note: when both athletes and judges take these oaths, they will be standing on wild boars' genitals.

Later in the afternoon, wander down to the ECHO COLONNADE, where the KERYX—the competition for heralds—and the SALPINKTES—the competition for trumpeters—will be going on. Today is also good opportunity to watch the ATHLETES TRAIN at the PALAESTRA and GYMNASION, explore the more informal OLYMPIC VILLAGE in the meadows south of the Altis, and to see whether you can crash any of the more upmarket PARTIES in one of the grand villas.

DAY TWO: HORSE RACING

DAY TWO begins with a procession. The priests of Zeus and the Hellanodikai will gather at the PRYTANEION, light up torches from the eternal flame, and visit dozens of the many small altars to the god that are scattered around the Altis. There will be BLOOD SACRIFICES conducted at many of these, so do expect a lot of flies. Once this process is completed, it will be time to get to the HIPPODROME.

Horse racing is a rich man's game. It is no mean financial and logistical feat to get your rides to this obscure corner of the Peloponnese from across the Roman Empire. The oldest and most prestigious of the three types of racing is the FOUR-HORSE CHARIOT RACE. There will also be TWO-HORSE CHARIOT RACES, originally contested with lowly mules, and simple HORSE AND RIDER RACES.

The chariot races are contested by teams of adult horses, who do twelve circuits of the Hippodrome, and by foals, who do eight. There may up to forty teams in any one race, making the hairpin turns exciting and perilous. The horse-and-rider event is notable for that absence of saddles and stirrups and thus the ever-present danger of jockeys being dismounted. Whoever wins, it won't be the rider who gets the laurels and the glory; this goes to the owners of the teams.

Do take note of the HIPPAPHESIS—the igneous starting mechanism that has been devised for these races. This will release the horses on the outside of the track before those on the inside of the track, which gives a running- start advantage to the outside lanes, compensated for by the shorter distance faced by the inside lanes. The ropes that secure the horses in their starting gates are held by a bronze eagle in a small altar in the centre of the track and bronze dolphins at the other ends.

DAY THREE: THE PENTATHLON

Today is given over to the PENTATHLON—the fivefold challenge—beginning with RUNNING, followed by DISCUS, JAVELIN, and JUMPING, and concluded with bouts of WRESTLING. The running race will be a sprint, one length of the stadium.

The DISCUS, which began life at the Olympics as a stone and then a bronze disc, will now be made of iron, and subject to the peculiar rule that the contestant who brings the heaviest disc will have it used by everyone. Don't expect the kind of balletic spin and heave that you will have seen at modern Olympics. Here, the discus is thrown from a fixed position, the athlete standing on a small podium or BALBIS. Distances will be marked out by the judges using wooden pegs. Do bear in mind there are no safety nets at Olympia and the average weight of the discuss coming your way is five pounds.

The JAVELIN, also thrown for distance, is a light wood shaft, tipped with metal. Most contestants will also have a long, leather string wrapped around the javelin, which uncoils and imparts spin and control when thrown. This is the event most closely associated with music, and may well be accompanied by DOUBLE-PIPERS.

The JUMPING COMPETITIONS are more opaque. Contestants take a series of jumps from standing starts on a wooden board, but swinging a variety of weight and dumbbells to aid their take-off.

WRESTLING takes two forms at the Olympics: sand wrestling, or KATO PALE, a grasping, rolling-in-the-sandpit struggle; and the more formal standing wrestling, or ORTHOS PALE, in which the fighters stay on their feet, embrace in a clinch, and look for opportunities to trip or throw their opponent. The Romans, always more extreme in their violence, have introduced knuckledusters to some bouts. The referees will be keeping control with the use of a wooden rod, though there are few constraints. The breaking of fingers, however, is expressly prohibited, as is gouging.

Note that if an athlete has already won three of the first four events in the pentathlon, he will be declared the winner without the need for a final wrestling decider.

FOUL! THIS VASE SCENE OF A PANKRATION COMBAT SHOWS A FIGHTER TRYING TO GOUGE HIS OPPONENT'S EYE, AND THE UMPIRE ABOUT TO DISQUALIFY HIM.

DAY FOUR: THE FESTIVAL OF PELOPS

Today is the FESTIVAL OF PELOPS, an intriguing and complex cult figure at Olympia. As a boy he was cut up and cooked by his father Tantalos and served to the gods—who were not fooled. The boy was reconstructed from the meal by the divines and he went on to win the hand of Hippodamia, daughter of Oinomaos, King of Pisa, in a race against her father's chariots—themselves gifts from the gods. Pelops only won the race because he bribed the opposing rider—a course that saw tragedy descend on him and his descendants. Nonetheless, by the late sixth century BC he had become a deity in Olympia, where he was buried.

The entire Hellanodikai, as well as the many hundreds of ambassadors attending from distant cities and Roman provinces, will all join the procession to the ALTAR OF ZEUS, where they will sacrifice and roast a hundred oxen. It's a cue for feasting. Once you've filled up on beef, take a stroll back to the stadium, where BOYS' DAY will be under way, with Olympic competitions reserved for the youth.

DAY FIVE: RACING AND COMBAT

The day begins with the FOOT RACES, which will be contested over three distances: the DOLICHOS, literally the long one, a twenty-four-circuit race nearly three miles long; then the STADE, a single dash of 210 yards; and finally the DIAULOS, twice the length, up and down the stadium with a single hairpin turn. There may be up to twenty-two competitors in each race and, depending on the numbers attending the Games, there may well be heats as well as final races. As the runners arrive, you should be able to see them position their feet in specially designed stone grooves built into the track that work as starting blocks.

The races will be followed by COMBAT SPORTS. First WRESTLING, conducted along the same lines as the pentathlon, and then BOXING. Recognisable to contemporary eyes, the fighting is much fiercer and rougher than now. Don't be surprised to see bare fists wrapped in leather thongs and occasionally embellished with studs. Next up is the PANKRATION, the "all-power thing." There really are no rules here, apart from no biting and no gouging, so expect ugly scrapping till somebody gives in or passes out.

The day will end with the HOPLITODROMOS, a race in armour where the otherwise-naked contestants don helmets and carry a small round shield. In the Hellenic era, athletes also wore greaves, but these have been abandoned, much to the chagrin of some.

DAY SIX: FESTIVITIES AND CHAMPIONS

On the final day, there will be a PROCESSION OF ALL THE OLYMPIC CHAMPIONS to the TEMPLE OF ZEUS, where they will receive their crown of olive leaves. Along the route the crowd will gather to shower them with fruit, twigs, and flowers. If you can't get into the temple—it will be crowded—do hang around outside to hear a rendition of the OLYMPIC HYMN sung to the champions.

A special banquet will then be held by the Hellanodikai for champions in the Prytaneion back in the Altis. There is no point trying to get yourself an invite to this occasion. In any case, you will note that the crowd is beginning to thin as people head for river, boats, and carriages. Lucian, a visitor to the next Olympiad in 165, found himself caught out by the crowds: "The end of the Olympic Games soon came—the best Olympics which I have seen, of the four which I have attended. It was not easy to get a carriage since so many were leaving at the same time, and therefore I stayed on for another day against my will."

Best not to tarry, and to make your way to the DEPARTURE POINT at the hotel complex by the river.

Opening Night at Shakespeare's Globe

JUNE 11–12, 1599 ✵ LONDON

THE RECONSTRUCTED GLOBE THEATER on London's South Bank gives contemporary audiences a flavour of what it might have been like to see the work of the world's greatest playwright in its original setting. But why not see what it was *actually* like? On this trip our travelers will attend the very first play ever staged there—the PREMIERE OF *JULIUS CAESAR*. Amidst the electric atmosphere of an Elizabethan theater, you will enjoy the excitement, thrills, and spectacle of this historic moment, and hear some of the most quoted lines ever written uttered for the very first time.

While theater today is an eminently respectable art form, back in Shakespeare's day it was the target of much hostility and criticism. Accommodating the poor alongside the well-to-do, thereby erasing the social distinctions and hierarchies that defined and ordered the Elizabethan world, theaters were an egalitarian force, a shared ritual that offered not only entertainment but a running commentary on the state of the nation. As a consequence, they were constantly under attack from religious figures, and London's governing bodies were always looking for an excuse to close them down, and periodically did so. However, nothing could stem the public's appetite for drama, and it was the genius of Shakespeare—and his brilliant contemporaries, Christopher Marlowe and Ben Jonson—that helped elevate its status. For the first time, playwrights were named and credited for their work, and plays began to appear in print alongside poetry and prose. Though, for many, the theater would never be respectable, it was institutions like the Globe that secured the legacy of Shakespeare and his fellow scribes, ensuring their words would survive to inspire future generations.

This thirty-six-hour trip will also include a day's sightseeing in London, allowing you to experience this great metropolis—the largest in Europe at the time—as it inspired and nurtured Shakespeare, the country boy with stars in his eyes. The Bard clearly fed off its cosmopolitan energy and the rapid social change affecting the city—and England—as the medieval world faded away and a more dynamic one came into being.

BRIEFING: THE BIRTH OF THE GLOBE

Around dawn on December 28, 1598, a dozen armed men trudged through the snow towards the site of The Theater Playhouse (or simply The Theater) in Shoreditch. Built in 1576, this was one of the oldest and most renowned theaters in London and was home to the Chamberlain's Men, a company of actors formed in 1594 under the patronage of Lord Hunsdon, the Lord Chamberlain, which included William Shakespeare, Richard Burbage, Will Kempe, John Heminges, Augustine Phillips, and Thomas Pope.

Having performed more than a hundred plays at The Theater, the Chamberlain's Men had a problem: they had failed to renegotiate the lease with the landlord, Giles Allen, while a plan to open a new venue in Blackfriars had fallen foul of its well-heeled residents. An alternative site had been found south of the river in Southwark—beyond city limits and the long arm of the Privy Council—near the ROSE THEATER, used by their main rivals the ADMIRAL'S MEN. This was secured on a thirty-one-year lease to run from Christmas Day 1598. However, the company still needed to construct an actual theater. Starting from scratch would be ruinously expensive. So they came up with an audacious scheme. Though Allen owned the land The Theater stood on, the Chamberlain's Men owned the building. Why not dismantle the whole thing and transport it lock, stock, and barrel to the new site?

With Allen on holiday in Essex and a royal performance pencilled in for New Year's Eve, time was of the essence. Having faced down some of Allen's friends, the men went to work, supervised by the master builder Peter Street. By dusk they had removed the whole frame of The Theater, with its huge, heavy square posts, and stored it in a warehouse by the river. Four days later, the job was done, and the materials were ready to shift by boat and cart once the foundations of the Globe were dug. (Allen later sued the Chamberlain's Men but lost the two-year court case).

This part of the process was near completion by early spring 1599, but then a cold snap slowed construction work and was followed by heavy rain and flooding at the end of May. However, once the weather cleared, the foundations were finished and the carpenters and sawyers moved in to fit the reassembled frame together. The outer walls were hoisted into place, and cross-frames and curved braces added for stability. Then "setting up" could start: with fresh timber arriving every day, new joists, rafters, partitions, seating, and staircases were built. Glaziers appeared, as did plumbers, smiths, thatchers (for the roof), plasterers (who used lathe, lime, and hair to cover the wood frame to give it the appearance of stone), and painters and specialists to marble the wooden columns on stage.

As summer approached, the new theater—named the Globe—was nearing completion. Saturday, June 12th, was chosen for the grand opening. The date (on the Tudors' Julian Calendar) coincided with the SUMMER SOLSTICE, a full moon, and the appearance of Venus and Jupiter in the night sky, all signs that augured well for the theater's future. For a society obsessed with astrology, such auspicious portents could not be ignored.

All the Chamberlain's Men needed was a new play to open with. Shakespeare, who had recently moved home to Southwark, close to the old Clink Prison, had already written *Henry V* that year, and now turned his attention to the classical world. He quickly knocked out *Julius Caesar*—2,500 lines total, almost all verse—had it approved for performance by the chief censor, the Master of the Revels, and rehearsals began. At these, each actor learned his part from individual scrolls, with only his bits of the text written on it—the whole company working flat out to be ready for the premiere.

THE TRIP

You will arrive at 9:30am on Friday, June 11, 1559, in the fields that straddle the fringes of urban SOUTHWARK. Bear due east for about a quarter of a mile and you will hit the main route to LONDON BRIDGE, the entry point for every kind of visitor from the continent and from the grand estates, farms, and villages of southeast England.

DRESS, ROOMS, AND VICTUALS

DRESS CODE in the Elizabethan era is used to distinguish social rank or profession. You will be wearing items that place you amongst the middling sort; not gentry or nobility, but still affluent and successful, borne aloft by self-made money, much like Shakespeare. Current trends are for colorful, eclectic, European styles: Spanish sleeves, French gowns, Dutch cloaks.

As the weather will be pleasantly warm, you won't be needing too many layers of clothing, but WOMEN TRAVELERS will require a number of different garments under their dresses (bared arms and legs are out, but generous displays of cleavage are OK if you're unmarried). To cover your delicate parts, you will have a length of washable linen, over which will be a chemise, then a linen smock with foreparts laid over it, and around that a fabric-covered framework (*farthingale*) to support your petticoat. Finally, you can put your gown on, with its ruffs and ruffles to add a touch of class, a neckerchief (*partlet*) and optional *mantles* to decorate the shoulders. You will have silk stockings and, considering the amount of walking you are likely to do, soft leather slippers with buckles. Extra adornments will include jewellery and a wig. To combat the sheer stench of the city, you will be wearing strong perfume—*amberg*, made from whale sperm, or *civet* from the cat of the same name—and carry a linen pouch stuffed with lavender.

MEN will have less clothing to negotiate: a linen shirt with a doublet and waistcoat, a cloak for theatrical effect, hose, breeches, and high leather boots. A felt hat with a colored hat band and dyed feather, an earring, and dagger worn at the waist complete the ensemble. You may fancy one of the latest hairstyles imported from the continent, while a neatly trimmed beard is the fashionable choice.

ROOMS

The standard of ACCOMMODATION in London's taverns is quite high. Your room will be spacious and comfortable, its walls decorated with painted cloths. There will be a large four-poster, curtained bed with a feather mattress on top and feather pillows and pillowcases to rest your head on. There will be clean linen sheets, blankets, coverlets, and quilts. Men will don a NIGHTSHIRT and CAP, while women will pull on a smock-like NIGHT RAIL. Additional items include a candle, candle holder, and a chamberpot by the bed, or, if you're lucky, a CLOSE STOOL: a square box with a hole in the top and a padded seat to plant your behind on. You are best off using the facilities in your room, as the public latrines (*jakes*) are pretty foul. If you are caught short when you're out and about, there are three communal privies on Tower Street, while the largest is on London Bridge and empties directly into the Thames.

VICTUALS

Those hoping to gorge on a Tudor roast will be disappointed. To encourage people to consume more fish, it is against the law to eat meat on Fridays and Saturdays. Expect your FISH to be poached or boiled in heavily spiced sauces, with added fruits and natural sugars, served with vegetables and

PIPE AND A PINT AT AN ELIZABETHAN TAVERN.

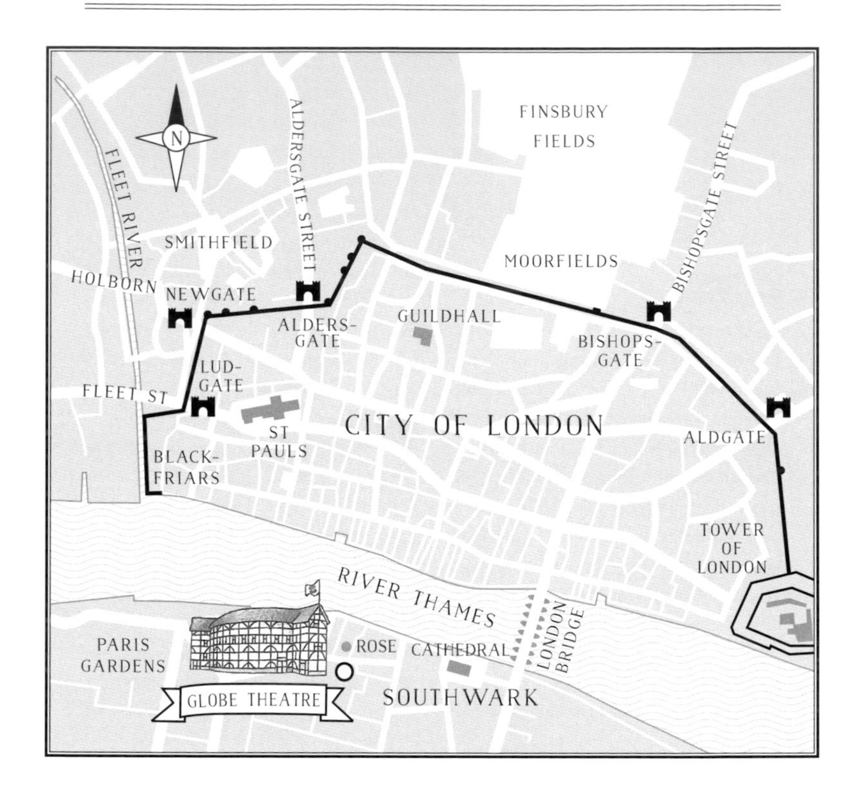

a salad (*sallat*) option, followed by cheese and dessert.

Most of your food will be sourced from one of London's many MARKETS, which you will encounter everywhere you go. Open six days a week, from 6am to 11am and 1pm to 5pm, they are all under the control of the Lord Mayor and his councillors, who fix prices and levy fines for offences like forestalling (intercepting goods before they reach the market). Your fish will come from Stocks Market, Billingsgate, and the fishmongers of Fish Street, your FRUIT AND VEGETABLES from Greenchurch Street Market or Queenshithe by the river (or any of the city's 377 grocers), while CHEESE will come from one of these markets or the cheese shops of Bread Street. Though you won't be able to eat any meat, you'll see plenty of it, whether alive and kicking at Smithfield or straight out of the slaughterhouses at Cheapside, where there are carcasses and joints hanging from hooks and skewered on spits.

Nobody drinks water. Instead, it's BOTTLED BEER, made from malt barley, water, and hops; ALE, which has no hops and a three-day shelf life; or WINE, of which there are thirty named brands available.

FRIDAY, JUNE 11TH: THE CITY

Jammed with traffic, human and animal, and lined with impressive buildings on either side, LONDON BRIDGE offers visitors a spectacular introduction to the CITY OF LONDON. Immediately when you step foot on it you will see two corn-grinding mills, then a drawbridge (no longer functioning) with a tower that sports the skulls of sixteen executed traitors and, midway across, a disused chapel.

Interspersed between these landmarks are large four-storey dwellings belonging to wealthy merchants, some with shops on the ground floor. The most grandiose of these is NONSUCH HOUSE, a Renaissance extravaganza, pre-fabricated in Holland and erected on site, with turrets, gilded columns, and carved galleries projecting out over the river. As you approach the northern end you will be confronted with an arch containing a massive water wheel. Passing this you will now be close to ALDGATE, the city's front door.

Criss-crossed by broad thoroughfares and linked by narrow streets and twisting alleyways, the CITY is a hotchpotch of dwellings, from the slums of the dirt-poor to the mansions of the filthy rich, all rammed together alongside official buildings, trading houses, retail outlets, and churches—many of them abandoned. There are many gardens and, always within walking distance, open green spaces, though many of these are occupied by laundresses drying their washing.

Wandering the streets, some paved or cobbled, others covered with gravel and beaten earth, you will notice how young everybody looks; half the population is under twenty. You will hear people being referred to as *Goodman This* or *Goodwife That*, and if somebody calls you *snout fair*, it means they think you are good looking. If, by contrast, they consider you *as much worth as a piss in the Thames*, take that as the insult it is clearly intended to be. Be wary of COURTESY MEN who prey on visitors, befriending then robbing them, while male travelers should be on their guard against A DEMANDER OF GLIMMER, an attractive women who promises you her body, takes your money, and runs. You may also find yourself being accosted by TOURIST GUIDES, ready to lead you a merry dance around town and relieve you of your cash. Don't be suckered into accepting their help.

You will meet SELLERS *crying their wares*, flogging *timely* (fresh) fruit and pies from baskets, BUSKERS, STREET PERFORMERS, and BALLAD SELLERS; at some point you will probably be handed a playbill advertising the grand opening of the Globe. BEGGARS will be hard to avoid—they are all licensed by church authorities to operate in a particular area for a specific period of time. You may also come across people being paraded around on CARTS (*carting*); these are minor offenders being publicly humiliated for their misdemeanours. Some may be tied to horses, but facing backwards. You will see individuals with either one ear or both ears missing: these are ex-cons who have been subject to a brutal punishment called *pillioring*.

LUNCH IN BISHOPSGATE

The main meal of the day is at midday. We recommend you dine at either THE BULL or THE GREEN DRAGON in Bishopsgate. You will be expected to wash your hands before eating and will be given a towel to dry them with. The menu will not be as elaborate or extravagant as some of the more extreme examples of Elizabethan cuisine. Nevertheless, you will be able to choose dishes like ELUS BAKYN, eels baked in red-wine sauce; PYKES IN BRASEY, grilled fish in wine; or SALMON poached in beer, vinegar, and herbs. CHEESES on offer will include green (new) cheese, hard cheese (cheddar), or soft cheese with herbs. DESSERT will come in the form of a fruit tart, perhaps filled with strawberries in red wine with sugar, cinnamon, and ginger, or cherries with mustard, cinnamon, and ginger. Or you might feel like a TRIFLE made with a pint of thick cream, seasoned with sugar and ginger and rosewater.

The best BEER to drink is *March Beer*. If you have an aversion to sweet varieties of WINE, then avoid the ever-popular *Sack*, a dry, amber wine from Spain that has added sugar; *Malmsey*, from Crete; *Muscatel*, from France; and *Rumney*, from the Balkans. For lovers of a full-bodied red, there is *Bastard* from Burgundy or, for a lighter bouquet, *Claret* from Gascony. The premium white wines are from La Rochelle or the Rhineland. After the meal, you may want to order a shot of Flemish BRANDY (*brandewijn*).

AFTERNOON SPORTS AND THE TOWER

Eating done, you can plunge back into the hurly-burly of the city, or you might prefer something a bit more recreational. If you fancy a spot of SWORD PLAY, we can book you a fencing lesson at Ely Place in Holborn. Beginners (*scholars*) will get instruction in the rapier, quarterstaff, and broadsword. Your teacher will be a member of the Company of Masters of the Art of Self-Defence. If PISTOL SHOOTING is more your thing, then head to ARTILLERY YARD, just to the east of Bishopsgate, an enclosed space with brick walls and targets at each end. Or, if ARCHERY appeals, make your way north out of the city to FINSBURY FIELDS, where there are 200 *butts* (targets) available for experts and novices alike.

Fans of the GRUESOME AND GROTESQUE might take a guided tour of the TOWER OF LONDON. However, do bear in mind that its dungeons and torture chambers are not just for show: they are very much in demand. Once inside, you will see THE PIT, a twenty-foot-deep hole where prisoners rot in total darkness; the infamous RACK, where limbs are stretched and bones broken; and THE LITTLE EASE, a cave too small to stand upright in. You will also be introduced to a particularly fiendish contraption known as the *Scavenger's Daughter*, a multifaceted instrument that comprises an iron band to compress victims' feet and head into a circle, iron gauntlets to crush their hands, and *jetters* (irons) to encase their arms and ankles.

EVENING PURSUITS

You will be staying the night at THE BELL INN by St Paul's Cathedral—the original Norman version of 1087. A LIGHT SUPPER will be served at 6pm, with a musical accompaniment provided by fiddlers, bagpipers, and ballad singers. The atmosphere will be very smoky; you can add to the tobacco haze by purchasing a small-bowled PIPE. Those with a competitive streak may fancy a game of CHESS or a crack at the BOWLING ALEE. There will also be GAMBLING going on around you, whether on card games—*gleek*, *primero*, *one and thirty*, *new cut*, and *trumps*—or dice, which are almost always loaded. Don't get involved unless you are very skilled with a blade; disputes over money and honour can escalate rapidly into duels to the death.

Whatever you do, you must be off the streets by the 9PM CURFEW, which is signalled by the ringing of church bells. Watches, with powers of arrest, patrol the city to ensure that the taverns are shut, the shops shuttered, and all good citizens are safely behind closed doors.

SATURDAY, JUNE 12TH: SOUTHWARK

The City Gates reopen at dawn. BREAKFAST will follow shortly after: a bottle of beer and buttered white or brown bread, each loaf sealed so the baker can be identified, the price and weight set by the Lord Mayor. After breakfast you will have a couple of hours to kill before having to be at Blackfriars to catch a WHERRY (water taxi) over to Southwark. You may want to take a look inside ST PAUL'S. Though famed for its beauty, the cathedral is now in a pretty sorry state, having suffered the depredations of the Reformation—its icons, tapestries, sculptures, and gold ornaments have all been ripped out. Nevertheless, it retains the awesome architectural scale and majesty we associate with Gothic churches. If you are feeling energetic, why not climb its 300 steps to the roof, from where you will get uninterrupted views of London.

You may also want to browse the BOOKSHOPS congregated round St Paul's Churchyard, selling editions in ancient Greek, Latin, Italian, French, and English; top of the bestseller list is *Foxes Book of Martyrs* (1563). Also popular are almanacs, saucy Italian verse, and home-grown plays, including some of Shakespeare's. (We regret that you may not make any purchases.) It is also worth a walk to nearby St Paul's Cross, where you will be harangued by fiery EVANGELICAL PREACHERS, radical Protestants with a mission to remind you of your sins and the dreadful ordeals awaiting you in the afterlife.

Once you are at BLACKFRIARS you will have no trouble finding a WHERRY. There are hundreds working the river, each one with two upholstered seats at the back, a canopy to shield you from the elements, and an oarsman up front who has done a two-year apprenticeship and been certified as competent by the Eight Rulers (overseers). The journey will cost you a penny.

As you cross the Thames you will be struck by the sheer number of vessels afloat. Aside from the wherries, there will be larger passenger barges with ten oarsmen, long tilt boats with a separate steerage boat attached, and dozens of large ships queuing up to unload their cargoes at the custom houses by London Bridge. Masses of fish inhabit the river's murky grey waters, while swans glide over its surface.

AROUND THE GLOBE

Southwark stretches south for a mile from the riverbank, and is heavily built up. Its squalid tenements are crammed together along packed streets and are home to watermen, craftsmen, and foreigners, who rub shoulders with an underworld of criminals and prostitutes. Many of Southwark's 300 inns double as BROTHELS, the most well known being the CARDINAL'S HAT. Mixed in with all the usual frightful odours is the unpleasant smell of the *stink* trades of brewing and tanning.

Your best bet for LUNCH is the ELEPHANT TAVERN on Horseshoe Alley, right next to the Globe and only a short walk from where your wherry docks. Don't expect fine dining: this is Southwark, and you're here for the theater. To guarantee your place for the 3pm performance, be there at least an hour early. As the GLOBE—a hundred feet in diameter, with a flag flying from the roof and the motto *Totus mundus agit histrionem* (The whole world plays the actor) inscribed above the entrance—has a 3,300 capacity, the crowds gathering outside will swell considerably as you wait to go in. You will see a few noblemen on horseback, some on foot, well-dressed members of the gentry, and gaggles of law students from the Temple Bar and Inns of Court mingling with tradesmen, apprentices, and labourers.

There will be CONJURORS AND JUGGLERS dazzling you with their skills, sellers offering refreshments: oranges, apples, nuts, gingerbread, and bottled beer, as well as pipes and pouches of tobacco at threepence a pop. Be on the alert for PICKPOCKETS and PROSTITUTES on the prowl. The queuing arrangements will be more disorderly than orderly until you reach the GATHERERS, who will take your money at the door.

Once inside the AUDITORIUM you will be impressed by the gilded decor, rich with classical motifs expressed in paintings, sculptures, tapestries, and hangings. Here you will turn either left or right to enter one of the three galleries lined with wooden benches that surround the fifty-foot-wide stage (2 pence admission). Be prepared to fight for a good seat, as there are roughly 1,000 other playgoers who want one too. If you want to rough it, you could pay just a penny to enter the seatless PIT, an open yard that slopes towards the stage, covered with ash, slag, and hazelnut shells, crammed with over 2,000 people from the lower sort—the *understanders*—jostling for a good view of the action.

While you will be exposed to the full glare of the June sunshine, the five-foot-high stage will remain totally in the shade. You will immediately notice how bare and functional the performance space is. There are two exits/entrances on either side;

between them is a curtained area used for DISCOVERY SCENES, which might feature a character asleep in bed or in his or her death throes, while a canopy is supported by two wooden pillars.

Otherwise, aside from a few props—tables, chairs, etc., that are kept under the stage (*Hell*)—the boards are bare. There is a trapdoor, but it won't be needed this afternoon. Just above the stage is a balcony that houses half a dozen MUSICIANS armed with trumpets, drums, other horns, recorders, and lutes. Torches and candles provide whatever on-stage lighting is required, and rudimentary sound effects, such as horses' hooves, birdsong, and bells, are supplied by the BACKSTAGE TEAM comprising a bookkeeper, assistant stage keeper, a carpenter, and two further stagehands.

Throughout, the audience will be fully engaged with the action. Cries of *mew* indicate displeasure. Expect the galleries to rise to their feet during the most intense scenes, while from the Pit there will be boos, hisses, shouts of encouragement and cheering, and applause for memorable speeches. Though rowdy, the punters are generally attentive and appreciative; some even bring table-books with them to note down significant passages from the play.

THE PLAY'S THE THING

Julius Caesar will last just over two hours, with no act breaks; the scenes, seventeen in all, will simply flow together to give the impression of continuous movement. Of the sixteen actors in the company—six of them boys who will take on the female roles—the leading man is RICHARD BURBAGE, the best of his day, and a pioneer of a more naturalistic approach to his craft.

Generally, the style of acting will be what we would call *hammy*. This is partly to compensate for the fact that much of the audience will struggle to see the performers' faces, while any dialogue spoken quietly will simply be lost in all the crowd noise. As a result, voices are big and declamatory, the verse projected with a booming rhetorical flourish. Gestures will be equally theatrical; the actors have a repertoire of fifty-nine hand gestures

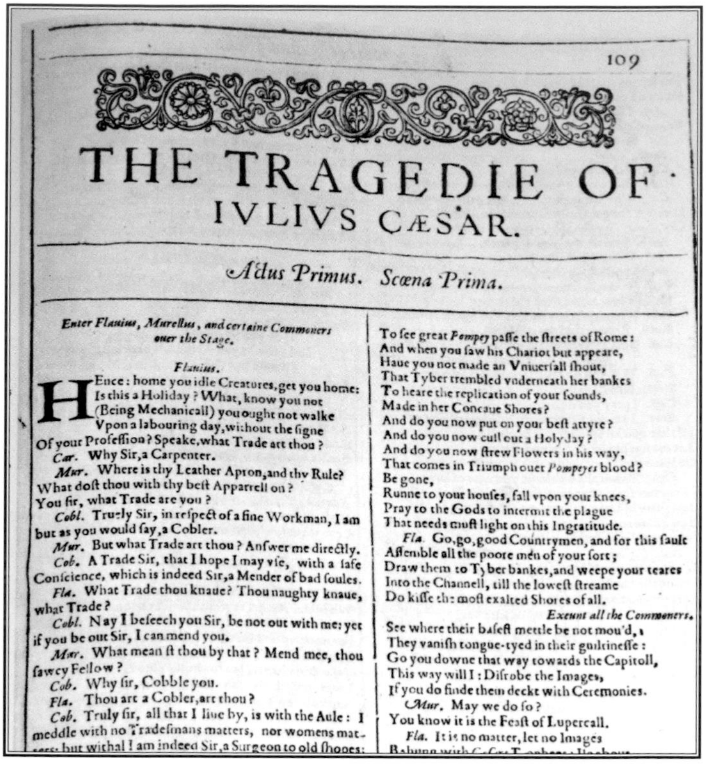
109

THE TRAGEDIE OF IVLIVS CÆSAR.

Actus Primus. Scœna Prima.

Enter Flauius, Murellus, and certaine Commoners ouer the Stage.

Flauius.

HEnce: home you idle Creatures, get you home:
Is this a Holiday? What, know you not
(Being Mechanicall) you ought not walke
Vpon a labouring day, without the ſigne
Of your Profeſſion? Speake, what Trade art thou?

Car. Why Sir, a Carpenter.

Mur. Where is thy Leather Apron, and thy Rule?
What doſt thou with thy beſt Apparrell on?
You ſir, what Trade are you?

Cobl. Truely Sir, in reſpect of a fine Workman, I am but as you would ſay, a Cobler.

Mur. But what Trade art thou? Anſwer me directly.

Cob. A Trade Sir, that I hope I may vſe, with a ſafe Conſcience, which is indeed Sir, a Mender of bad ſoules.

Fla. What Trade thou knaue? Thou naughty knaue, what Trade?

Cobl. Nay I beſeech you Sir, be not out with me: yet if you be out Sir, I can mend you.

Mur. What mean'ſt thou by that? Mend mee, thou ſawcy Fellow?

Cob. Why ſir, Cobble you.

Fla. Thou art a Cobler, art thou?

Cob. Truly ſir, all that I liue by, is with the Aule: I meddle with no Tradeſmans matters, nor womens matters; but withal I am indeed Sir, a Surgeon to old ſhooes:

To ſee great *Pompey* paſſe the ſtreets of Rome:
And when you ſaw his Chariot but appeare,
Haue you not made an Vniuerſall ſhout,
That Tyber trembled vnderneath her bankes
To heare the replication of your ſounds,
Made in her Concaue Shores?
And do you now put on your beſt attyre?
And do you now cull out a Holyday?
And do you now ſtrew Flowers in his way,
That comes in Triumph ouer *Pompeyes* blood?
Be gone,
Runne to your houſes, fall vpon your knees,
Pray to the Gods to intermit the plague
That needs muſt light on this Ingratitude.

Fla. Go, go, good Countrymen, and for this fault
Aſſemble all the poore men of your ſort;
Draw them to Tyber bankes, and weepe your teares
Into the Channell, till the loweſt ſtreame
Do kiſſe the moſt exalted Shores of all.

Exeunt all the Commoners.

See where their baſeſt mettle be not mou'd,
They vaniſh tongue-tyed in their guiltineſſe:
Go you downe that way towards the Capitoll,
This way will I: Diſrobe the Images,
If you do finde them deckt with Ceremonies.

Mur. May we do ſo?
You know it is the Feaſt of Lupercall.

Fla. It is no matter, let no Images

THE FIRST FOLIO OF *JULIUS CAESAR*—NOT YET PUBLISHED.

to communicate different emotions and states of mind. Though all the cast are adept at *thribbling* (improvisation), today they will stick to the script. The actor's costume is called his *shape*, while his place on stage is his *habitation*. Over the course of *Julius Caesar*, they will be wearing togas, tunics, cloaks, robes, and armour (for the battle scenes), all the responsibility of the wardrobe keeper.

Once *Julius Caesar* begins, there will be many HIGHLIGHTS to look out for. After a group of plebs set the play in motion with their disparaging comments about Caesar's ambitions to rule single-handedly, SCENE 2 will begin with blaring music from the balcony, heralding the entrance of the main characters and a throng of citizens. Night falls during SCENE 3, and the plot thickens as the would-be assassins make last-minute preparations for murder most foul, accompanied by a wild and unholy thunderstorm, courtesy of the sound-effect guys: a sheet of metal will be used to make the thunder, fireworks for the lightning, a canvas tied to a wheel to produce the wind, and dried pies rattled on tin to create the rain.

The intense noise of the storm continues into the next scene, where Caesar ponders his potential fate, and then onto the killing itself in SCENE 8. When each conspirator plunges his dagger into Caesar's stricken body, bladders of sheep's blood, strategically placed on the actor's body, will erupt, spilling their contents onto the stage. As Brutus delivers the final thrust, Caesar, dismayed by the treachery of his closest companion, will speak those three words of Latin that will echo down the ages: "Et tu, Brute?"

This coup de grâce is followed by the famous funeral orations of Brutus and Mark Antony; both actors will ascend to the musicians' balcony to deliver their speeches. Mark Antony's stirring tribute to Caesar, featuring the immortal opening lines, "Friends, Romans, countrymen, lend me your ears. I come to bury Caesar not to praise him," will turn the people of Rome against the assassins and an angry mob will scour the city for them, happening upon the hapless poet Cinna in SCENE 10: thinking he is someone else, they will pummel him to death (more sheep's blood).

The play is now entering its final third: Brutus and his fellow conspirator Cassius are with their army near the town of Philippi, preparing for battle with Mark Antony and Octavius. There will be a lot of drumming, trumpet blasts, and off-stage shouting to help create a military mood. During SCENE 12, the night before the decisive battle, the ghost of Caesar will visit Brutus in his tent and issue the chilling warning, "Thou shalt see me at Philippi." To give their meeting a spooky, supernatural quality, stagehands will place lights behind colored bottles; the shapes and shades shed by them will bathe the actors in unearthly hues.

The last four short, fast-paced scenes encompass the battle and its grisly aftermath (lots more sheep's blood). Fearing all is lost, first Cassius, then Brutus, will fall on their swords, leaving the victors, Mark Antony and Octavius, to bring proceedings to a close.

Before you file out, however, the next play will be announced, the actors will kneel together and offer a prayer to the queen, and then the whole ensemble will perform a boisterous and ribald JIG to send you merrily on your way.

POST-PLAY

It is customary after the performance to promenade through PARIS GARDENS, a mere stone's throw from the Globe. Here you will be able to buy alcohol and watch games of bowls and cards. You will also be close to its BEAR-BAITING ARENA. This hugely popular form of entertainment features star bears with names like George Stone, Harry Hunks, and Harry Tame. Costing a penny to stand in the stalls, or tuppence for a gallery seat, punters will be treated to the sight of the bear being led in on a chained leash and then tied to a stake before being viciously assaulted by hunting dogs, usually Great English mastiffs. These ferocious contests can last several hours, before the combatants are too exhausted or wounded to continue.

DEPARTURE

With dusk fast approaching, and the streets becoming more dangerous for outsiders, you should make your way southwest out of Southwark, past farmsteads, woods, and marshy ground. You will soon be at the fields where your journey began and from where you will depart.

The Golden Age of Hollywood

MAY 28–JUNE 26, 1923 ✵ LOS ANGELES, US

ALL US FILM BUFFS ARE FAMILIAR WITH Cecil B. DeMille's last great movie, the 1956 *The Ten Commandments* starring Charlton Heston. But how about his original 1923 version, which pretty much defined epic, as the most expensive film of Hollywood's Golden Age? On this trip you get to be an extra, and—as one of the Israelites—experience first-hand the mad extravagance and unreality of the film industry on a huge location shoot in the Californian desert, living in a tent city especially erected to cater for the thousands of people and animals involved in making the picture.

As a bonus, before you embark on this adventure, you will spend forty-eight hours in Los Angeles, exploring a city in its energetic infancy, emerging out of the wilderness and beginning to assume the shape and character of the sprawling metropolis it will become.

BRIEFING: THE TEN COMMANDMENTS

In 1923, the 41-year-old Cecil B. DeMille is a well-established Hollywood director, famed for his sharp social comedies. However, his most recent film, *Adam's Rib*, has been a critical and commercial flop, falling foul of the growing moral backlash against Jazz Age decadence that Prohibition set in motion.

DeMille realises he needs to change direction and, never shy of publicity, arranges for the *LA Times* to run a "Nationwide Ideas Contest" for the topic of his next movie, with a first prize of $1000. Though he doesn't guarantee the winning idea would become his next feature, it turns out the fates were smiling. A letter from one F.C. Nelson, a manufacturer of lubricating oil from Michigan, begins: "You cannot break the Ten Commandments—they will break you." DeMille is hooked. A committed Christian, he realises at once that here was one of the greatest stories ever told—just awaiting its Hollywood makeover. His confidence in the idea is further boosted by seven other contestants urging him to take on the Exodus.

Work on the script begins in early 1923, spearheaded by DeMille's creative partner, muse, and mistress, Jeanie MacPherson. She decides the film should run in two parts: the first a prologue telling the story of the Israelites and their escape from bondage. This will begin with the Nine Plagues that devastate the Pharaoh's realm and take the story to the point when Moses descends from Mount Sinai with the Word of God and curses his people for worshipping false idols. Part two will be a contemporary domestic drama of two brothers: one who believes in the Bible, one who doesn't.

DeMille works exclusively for Paramount, the largest and most powerful studio in the world, whose success is due to the partnership between Jesse L. Lasky and Adolph Zukor. Lasky is a theatrical impresario who handles talent—he discovered Mae West—while Zukor is the business brains. The two came together in 1911 (the same year Lasky took on DeMille, then a jobbing actor and playwright) and three years later set up Paramount to distribute their movies. By 1921 they are churning out over 100 films a year, distributed through their nationwide chain of 300 cinemas.

Lasky and Zukor's initial reaction to DeMille's plan is positive, and they approve a $750,000 budget. But Zukor soon becomes concerned at the spiralling cost (the finished film came in at $1.5m) and tries to pull the plug midway through the shoot. Undeterred, DeMille raises the cash himself to complete the movie, loaning $1m from the head of Kodak and A. P. Giannini of the Bank of America. His gamble pays off. *The Ten Commandments* will be a huge hit, grossing more than $6m.

THE TRIP

You will be arriving in LOS ANGELES at lunchtime on Monday May 28, 1923. You will find yourself in Pershing Square, in front of the magnificent and newly opened BILTMORE HOTEL, where a small suite has been booked in your name. Billed as "the biggest and most sumptuous hotel in the country west of Chicago," the Biltmore is a sight in itself, created in a mad, magpie assembly of architectural styles: a Moorish ceiling picked out in gold leaf mixing it with Spanish Baroque doorways, Italian chandeliers, pseudo Renaissance frescoes, and statues of Roman gods. In your suite, male travelers will find a LINEN SUIT, COTTON SHIRT, and a selection of TIES, while women can choose between a SUMMER DRESS AND CARDIGAN or a WHITE PLEATED SHIRT WITH SLACKS.

THE BILTMORE PROMOTIONAL MATERIAL DOESN'T MAKE MENTION OF IT, BUT IT IS HANDILY KITTED OUT WITH ITS OWN SPEAKEASY.

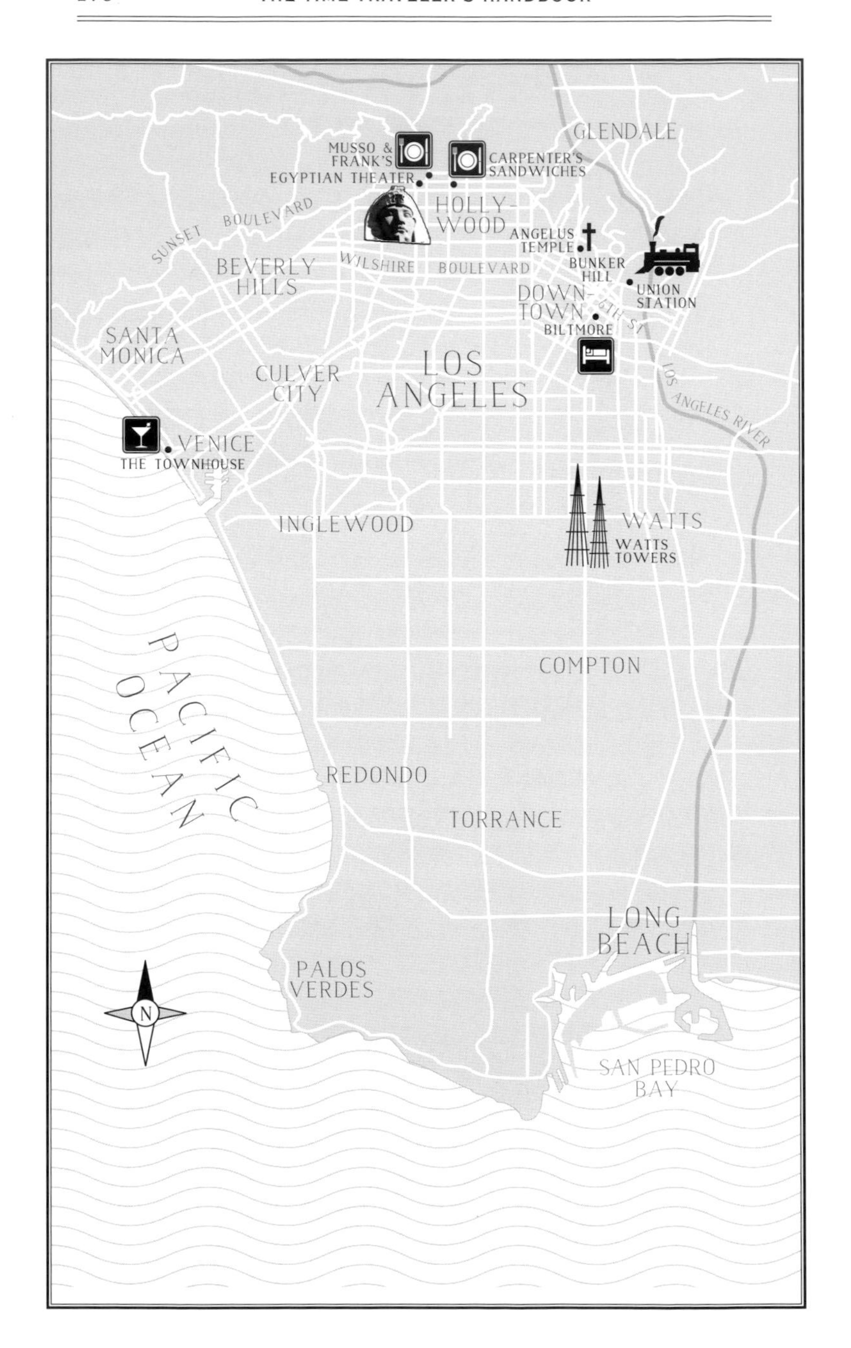
GLENDALE
MUSSO & FRANK'S
CARPENTER'S SANDWICHES
EGYPTIAN THEATER
HOLLY-WOOD
SUNSET BOULEVARD
ANGELUS TEMPLE
BEVERLY HILLS
WILSHIRE BOULEVARD
BUNKER HILL
UNION STATION
DOWN-TOWN
6TH ST
BILTMORE
SANTA MONICA
CULVER CITY
LOS ANGELES
LOS ANGELES RIVER
VENICE
THE TOWNHOUSE
INGLEWOOD
WATTS
WATTS TOWERS
PACIFIC OCEAN
COMPTON
REDONDO
TORRANCE
LONG BEACH
PALOS VERDES
N
SAN PEDRO BAY

Your main source of information about what's going on in the city will be the many daily and evening newspapers. Choose from the solidly establishment *Los Angeles Times* or more progressive *Los Angeles Herald Examiner* and its evening stablemate the *Los Angeles Express*, owned by Randolph Hearst. Keep your eye out, too, for the very lively Spanish-language press—*La Prensa* and *El Heraldo de México*—which is growing in tandem with the city's Latino population. RADIO has recently arrived in Los Angeles, too, with three stations currently on the air: KFI (640 AM), KHJ (720 AM) and KNX (1050 AM).

MONDAY-TUESDAY, MAY 28TH–29TH
EXPLORING LOS ANGELES

For most Angelenos the main forms of transport are the YELLOW CARS of the Los Angeles Railway (best for Downtown and nearby attractions) and the RED CARS of the Pacific Electric Company, which snake out across much of the city as far as Santa Monica, Long Beach, and the San Gabriel Valley. We do encourage travelers to use them, as by the 1950s they will have been closed down, ripped up, and removed, leaving the city bereft of rail transport for decades. You may also be able to hail a JITNEY, one of the unlicensed taxis operating in the city. As you could end up on the running boards of a Model T Ford, this option is not for the faint-hearted.

You are, in any event, at the heart of Downtown, which is the hub of the emerging city. BROADWAY is its main shopping street and over the past decade has been acquiring an incredible array of gargantuan PICTURE PALACES. It's worth buying tickets to whatever movies are on, just to experience these, chief among which are the MILLION DOLLAR, RIALTO, GLOBE, ARCADE and CAMEO.

BUNKER HILL, at the northwest corner of Downtown, will be unrecognisable to visitors familiar with LA's twenty-first-century incarnation as a cluster of futuristic skyscrapers. On this visit you will find it is still a residential neighbourhood, though many of its generously proportioned Victorian villas are being converted to bedsits and apartments and the whole area is acquiring a sleazier and more late-night feel than before the war.

Like today, very little of "Hollywood" is actually in Hollywood, with many of the studios located in Culver City and elsewhere. Nonetheless, the HOLLYWOOD district is certainly worth a visit, if only to visit the spanking new GRAUMAN'S EGYPTIAN THEATRE on Hollywood Boulevard. Any excursion to this district should also take in the intersection of Hollywood and Vine. Here you will get a superb view of the HOLLYWOODLAND sign. An advertisement for a real-estate project, it will eventually lose the "land," but for the moment you get to see the whole thing. Its lights are being installed and they will be switched on in July.

It is also worth making a detour to the ANGELUS TEMPLE (1100 Glendale Boulevard) in ECHO PARK, the city's first megachurch, built by its first super-populist evangelist Aimee Semple McPherson and her International Church of the Foursquare Gospel. Expect a big congregation, speaking in tongues, faith healing, and forthright requests for donations.

There is less to draw the visitor south of Downtown, but one trip worth considering is to the triangular plot at 1727 East 107th Street in WATTS, where new arrival Italian immigrant Simon Rodia has recently taken up residence. He is at the very earliest stages of building what will become known as the WATTS TOWERS, an extraordinary assemblage of Gaudi-like metal and wire spires decorated with fragments of tiles and bottles.

EATING IN LA

LA gastronomy is at a primitive stage. Aside from a few high-end hotels and secret Mexican *cantinas*, eating out is a solidly midwestern affair dominated by fried chicken; classic places like *The Brown Derby* are still a few years from opening. However, there are plenty of intriguing options, particularly in Downtown, where you'll find several branches of the PIG'N'WHISTLE (224, 452 and 712 South Broadway and 87 West 7th St). Originally confectionery stores with soda fountains, these have recently added seating and an extensive menu. Marvel at the amazing carved wooden ceilings, enjoy the fabulous tiling and the vast counters bursting with multicoloured candies.

If you fancy something more savory why not join the *French Dip* craze that is engulfing the area? Completely unknown in France, the French Dip sandwich is a lightly toasted soft roll piled with layers of roasted sliced meat and softened with a spoonful of jus or gravy. Two outlets claim to have invented it: COLES, a buffet on the ground floor of the Pacific Electric building (6th and Main); and PHILIPPE'S, which has recently relocated to just south of Union Station at 246 Aliso Street.

For a chance to mingle with highrollers, oil tycoons, and real-estate moguls, try the PACIFIC DINING CAR (1310 W 6th Street), an adapted railway carriage parked in a Downtown parking lot and serving up top-quality steaks and pies.

If you decide to venture farther afield in search of refreshments, head for CARPENTER'S SANDWICHES at the intersection of Sunset and Vine. With its fabulous neon-signed pagoda, this is one of the city's first authentic drive-ins that actually serves customers in their cars, though walk-ins are always welcome. So stroll in and sample the *Real Hamburger Sandwich* (just 15 cents), or the barbecue pork, sirloin steak, and fried oyster options. These can be washed down with *Ben Hur Delicious Drip Coffee* or an extensive list of sodas.

For a more upmarket dining experience, make your way to 6669 Hollywood Boulevard to a restaurant formerly called *Françoise's* which has just this year reopened as MUSSO & FRANK GRILL and is set to become a fixture in Hollywood high society. Waiters in bow ties and formal jackets, low lights, dark-wood fittings, and red velvet booths make for an intimate setting. Amongst the regulars, you may catch sight of Douglas Fairbanks, Rudolph Valentino (who is very partial to the spaghetti dishes), and Charlie Chaplin (who generally orders broiled lamb with capers).

Another haunt of Hollywood pioneers, like the young Walt Disney, is TAM O'SHANTER'S at 2980 Los Feliz Boulevard. The building has been designed by Harry Oliver, one of the leading art directors in Hollywood, in the "storybook style"—a movie-set take of fairy tale cottages with impossibly steep roofs and crooked windows. Inside you will find a riot of faux-medieval paraphernalia and Scottish tartan kitsch; the perfect backdrop to your steakhouse fare of prime rib and Yorkshire pudding.

GETTING A DRINK

It is 1923 and the 18th Amendment, prohibiting the sale and consumption of ALCOHOL, has been in force for three years. Of course getting a drink in Los Angeles is not a problem if you know where to go and who to ask. The city's police force, the LAPD, is probably the best informed about the whereabouts of the city's SPEAKEASIES as most of them are on the bootleggers' payrolls. Rather than approach an officer for advice, though, we prefer for you to stick to a few tried-and-tested locations where we can vouch for the quality of the product; it's important to note that there have been a considerable number of deaths already this year from locals consuming poisonous hooch.

Your most convenient locale is the GOLD ROOM, which can found right inside the Biltmore, where a certain Baron Long is known to organise gin-soaked sessions for hotel guests. Or, if you're curious about more regular locales, take the long ride from Downtown to Windward Avenue in Venice and ask for MENOTTI'S GROCERY STORE. Saunter in past the few boxes of tomatoes and herbs and you'll be directed to the back room, where the TOWNHOUSE will be in full swing at any time of day or night. Here you can be sure of getting the best-quality Canadian imported rums and whiskeys.

WEDNESDAY, MAY 30TH: ON LOCATION WITH CECIL B. DEMILLE

Having sampled what LA has to offer, you will be ready to begin the next stage of your trip: welcome to showbusiness. You will need to be at UNION STATION by 11pm on Tuesday night to catch the specially laid on trains that will transport you and your fellow 2,500 extras on a journey 170 miles north. Here you will find Cecil B. DeMille's *Ten Commandments* set, located beside the Nipomo dunes, outside of Guadalupe (population 1,130) in Santa Barbara County. Given the numbers battling to get on board the trains, it would be advisable to arrive at the station as early as possible.

CAMP DEMILLE

You will arrive around 4am on Wednesday morning at CAMP DEMILLE. This mini city covers 24 square miles and has 4 miles of sidewalk lining its streets (these are named after company executives—the main thoroughfare is Lasky Boulevard).

At the eastern end of town you will find the ADMINISTRATION TENTS, two huge mess tents (one of which has a projection room where each day's rushes will be viewed by DeMille and his intimate circle), storerooms, the cameramen's quarters, and a hospital manned by army surgeons. To the west, sheltered from the wind by looming sandhills, is the wardrobe HQ and prop rooms (24 tents in all), plus one large multipurpose tent which, each morning, will act as a schoolroom for the sixty-five kids on set.

Situated due north, and hard to miss thanks to the flag—deep blue with white letters spelling out Cecil B. DeMille's name fluttering above it—is the DIRECTOR'S MARQUEE, a palatial Oriental-style affair furnished with Persian rugs and other luxuries; outside of it is a garden of blue lupins installed at the cost of $1,700. DeMille's every need will be met by his personal staff of fifty-six PAs and domestics.

THE TEN COMMANDMENTS SET A NEW EPIC SCALE FOR HOLLYWOOD. AFTER THE MOVIE WRAPPED, DEMILLE ORDERED THE ENTIRE SET TO BE BURIED. THE SPHINX WAS EVENTUALLY REDISCOVERED SIXTY YEARS LATER.

Mingling with you and the other extras is the small army of WORKMEN AND ANCILLARY STAFF needed to maintain this vast enterprise: 500 carpenters, 400 painters, 380 decorators, and a similar number of landscape gardeners. There are also whole troops of electricians required to handle the camp's lighting (a portable electric plant with enough wattage to illuminate ten circus tents, and searchlights and arcs used for the night scenes), and dozens of telephone engineers to maintain 75 miles of cables and wires. Providing blow-by-blow coverage of proceedings is a coterie of journalists from publications like *Movietone News* and *Motion Picture News*. To service all these people, 132 trucks shuttle back and forth to Guadalupe carrying the camp's supplies, laundry, and garbage, at a cost of $40,000 a day.

You will also get accustomed to encountering all manner of ANIMALS, including 200 camels and scores of horses, mules, oxen, dogs, goats, chickens, geese, ducks, and guinea fowl, as well as a herd of cows to provide milk for the city. They are corralled in the lowlands just north of the camp, next to twenty-two tents occupied by their herders and animal trainers.

THE SET, designed by Paul Iribe, is constructed of 55,000 feet of lumber, 350 tons of plaster, and 25,000 pounds of nails. It was transported by lorry to the site on May 21 and re-creates the CITADEL OF RAMESES II, whose main gate (109 feet tall and 750 feet wide) has two colossal 35 foot-high statues of the Pharaoh either side, made with clay and plaster laid over metal frames. Its walls feature bas-reliefs of horse-drawn chariots, and it is approached along an avenue of 25 foot-high concrete sphinxes, each weighing around 4 tons.

ROOMS, FOOD, AND ENTERTAINMENT

You will be staying in one of the 1,000 or so tents, which are arranged in military style in two encampments, with Lasky Boulevard running between them. Your name will appear on boards at the end of your "street," with your exact address. In his infinite wisdom, DeMille has decided that living quarters will be segregated by sex; any of you traveling as a couple will be sleeping separately. This gender divide will be strictly enforced by DeMille's private police, both male and female units, who will zealously crack down on fornicators, bootleggers (Prohibition is in full force even here), and gamblers. Anybody caught indulging in these activities will be swiftly ejected from camp. So don't even think about smuggling any hooch or members of the opposite sex into your tent.

Each tent contains an electric light, two army cots, a bench, two washstands, two basins, a bucket, and a dresser. Also available on site are thirty shower baths for men and twenty-four for women; as you and all the extras will end each day's shooting covered in dirt and grime, these facilities are in constant demand and you will have to endure long queues if you want to use them.

There are two MESS TENTS, one for men and one for women, each with a 1,500 capacity and open 24 hours a day. The food is unexciting but plentiful, provided by 97 army cooks. If you are on set you will get a PACKED LUNCH: a sandwich (one of 7,500 delivered every day by wagons) and an apple or orange. Such is the scale of this catering job that three men are employed just to twist the necks of your sandwich bags.

To stave off boredom and restlessness, DeMille will lay on a variety of entertainment: movies, jazz bands, vaudeville, and circus acts, and weekly BOXING BOUTS. A highlight for fight fans will be the chance to see Fidel LaBarba, a dazzling flyweight who will win gold at the 1924 Olympics, get into the ring with Brian Moore (his little brother Pat is playing the Pharaoh's son) for the prize of a $20 gold piece. A true gentleman, LaBarba, despite his superiority, will let Brian win the contest.

As an extra you will be paid $10 per day. This remuneration may not be sufficient compensation for over three weeks on set without alcohol or sex to relieve the strain of performing in a series of draining and arduous scenes, or help pass the long periods of inactivity while being beset by grisly weather—thick fog rolling in off the Pacific and bitterly cold 60-mile-an-hour winds bringing with them stinging sand storms. To his credit, DeMille is completely honest about the challenges ahead: "You will miss the comforts of home, and you will be asked to endure perhaps the most unpleasant location in cinema history."

If after a week you've had enough of these testing conditions, an early departure time is available at midnight on June 8: make your way to the CITADEL ENTRANCE and you'll leave from there. However, if you choose this option, you will miss out on participating in some of the most memorable sequences ever committed to celluloid.

THE SHOOT: MAY 31ST–JUNE 8TH

Wearing a flannel shirt, airforce jacket, khaki trousers, puttees, and his signature knee-high boots (worn due to his weak ankles and chronic fear of snakes), DEMILLE will rule over the set with a rod of iron. An authoritarian perfectionist with an exceptional eye for detail, he will do whatever it takes to get the shot he wants: he will bully, plead, charm, and cajole everyone into giving 100 per cent to every scene. Thankfully, due to the vagaries of the climate and the sheer expense of proceedings, you will not be required to

CECIL B. DEMILLE ON SET WITH CHARLES DE ROCHE (PHARAOH) AND PAT MOORE (PHARAOH'S SON), ONE OF THE SIXTY-FIVE KIDS ON SET.

maintain these high standards for innumerable takes. However, you will have to be on your best behaviour, as DeMille, using his field glasses, will be watching each actor like a hawk, spotting anything untoward in the crowd: smiling, giggling, chewing gum will all result in a public bawling out.

To ensure he gets his footage as quickly and efficiently as possible, DeMille will deploy up to seven cameras simultaneously (each costing around $4,000). The maestro will have two operating next to him, one for long shots and one for closeups. He will use two assistant cameramen—Donald Biddle Keyes and Edward Curtis—and a stills photographer, Eugene Richee. His trusted lieutenant, Cullen "Hezi" Tate, will act as his assistant director running the second unit, while the look of the film will be created and managed by his cinematographers: Bert Glannon, J. Peverell Marley, Archie Stout, and Fred Westerberg. Working alongside them is Ray Rennahan who will be manning a technicolour camera provided by The Technicolour Company, which will use a two-colour system known as "subtractive" to imbue several sequences with a reddish tint.

THE STARS

The lead role of MOSES has been assigned to the sixty-two-year-old THEODORE ROBERTS, a veteran of twenty-three DeMille productions. Sporting flowing white hair and a waist-length beard, and carrying a long staff, he brings dignity, gravity, and passion to the part. The French actor CHARLES DE ROCHE will play the PHARAOH with a still, haunted expression in his eyes, while the main female roles are Pharaoh's wife—JULIA FAYE—and Moses's sister MIRIAM—ESTELLE TAYLOR.

EXTRA EXTRAS

After a few days' shooting you will notice the arrival of 250 Orthodox Jews, gathered from recent immigrants and elders among California's Orthodox community by RABBI AARON

THEODORE ROBERTS LEADS THE WAY AS MOSES, BUT DEMILLE INSISTS ON PASSIONATE COMMITMENT FROM EVERY ACTOR ON SET.

MARKOVITZ. They are here to bring added authenticity, and their appearance alone will be enough to conjure up visions of ancient times. Or, as one of the resident journalists described them, "old men, infirm of step, feeble with long hair and patriarchal beards. With their belongings tied up with newspaper or battered old suitcases, they huddled together."

Their first day on set is marred by controversy when they are given ham sandwiches for lunch. After the inevitable uproar, DeMille will have a kosher kitchen established for their benefit.

LIGHTS, CAMERA, ACTION!

A bugle will sound REVERIE at 4:30am. Rise and make your way to the end of your street, where there are call boards with details of the day's shoot. Then line up at the appropriate mess tent for breakfast. Once you have eaten, DeMille's assistants will organise you into groups known as COMPANIES or PLATOONS. Before you can get into costume, however, you must undergo a series of bizarre rituals. As you are meant to be in the North African desert, you must look like you are baking hot and covered in sweat. To achieve this effect, you will strip off and be sprayed with GLYCERINE from a giant tank holding 500 gallons of the stuff. To give the impression of sunburn and deep tans, you will then be covered from head to foot with special oils.

Transformation complete, head off to the WARDROBE AND MAKE-UP TENTS, where Claire West and Howard Greer are in charge of 16 miles of cloth, 3 tons of leather for the harnesses of the chariots, 2 tons of talcum powder to whiten the actor's faces (Rameses and his inner circle all wear a ghostly pallor), and 200 pounds of safety pins to hold everything together. Younger male travelers will be wearing LOINCLOTHS; some of the more athletic amongst you be wearing hardly anything. Older men will have FULL ROBES, as will most women, though the more photogenic females will be scantily clad. Some of you will wear sandals, others will be barefoot. To keep you warm between takes, you will be issued with a blanket.

Another bugle call will signal the START OF SHOOTING. Hay wagons will transport you to your location. DeMille, unlike other directors, prefers to shoot in dramatic sequence. Though you will not be in every crowd shot, just about every extra will be employed in the really big set pieces.

EXODUS

Your first major scene is the ISRAELITES LEAVING PHARAOH'S CITADEL. You and the rest of your tribe will be crammed into an area outside the gates demarcated by 20-foot-high parallels arranged on sand dunes about 100 feet away from the entrance. DeMille will fire a pistol and, accompanied by a TRUMPET FANFARE composed by Hugo Riesenfeld, you will start streaming out of the Citadel, a confused and scared mass of humanity dragging your meagre possessions with you, aided, and at times hindered, by mules, camels, and oxen, while flocks of sheep and herds of goats will add to the frenzied confusion all around.

Next, you will CROSS THE DESERT, buffeted by strong winds and assailed by billowing clouds of sand. As you trudge wearily along you may well get a sense of what it must have felt to be a refugee fleeing the Pharaoh's wrath; certainly, it has this effect on the Orthodox Jews, who will spontaneously begin to sing in Hebrew, "Father of Mercy" and "Hear O Israel the Lord Our God, the Lord Is One" with such emotion that their plaintive, soulful voices will reduce DeMille to tears.

CHARIOTS AWAY!

Though you are not directly involved in the CHARIOTS SEQUENCE, try and find a good vantage point, as this will be the most dramatic and dangerous action of the whole shoot. The RIDERS are drawn from the 11th Cavalry and the 2nd Battalion 76th Field Artillery, under the command of Lt Tony McAuliffe, and they don't hold back. Sixty of them are injured, four severely. The HORSES, black thoroughbreds brought in from Kansas City at a cost of $50,000, and marshalled by their trainer A.F. Stultzman, will not fare much better. Many are lamed and wounded, their flesh torn apart. Animal lovers may want to steer clear.

Wearing golden tunics and gilded helmets with multicoloured plumes, CHARIOTEERS will gather first outside the Citadel. Chaos reigns as 250 chariots try to manoeuvre in a small space. The horses, whipped into action, panic and bolt, tearing off in all directions, toppling and unseating their drivers. Caught in the mayhem are thirty members of the PALM COURT ORCHESTRA wearing tuxedos and evening gowns. Led by Rudolph Berliner, they will be playing martial music in a special enclosure off camera. Though wild horses will be stampeding right at them, the band will play on regardless until the runaway steeds crash into them, breaking instruments and injuring several musicians.

LIGHTS, ACTION, CHAOS! THE LEGENDARY CHARIOTS SEQUENCE KICKS OFF.

Order will be restored in the next sequence, as rows of chariots speed at full pelt across the flat desert plain, with the Pharaoh's two black stallions ($10,000 on the balance sheet) leading the charge. It is truly an awesome sight to behold. However, there is trouble ahead. When asked to negotiate getting down a very tall and steep set of sand dunes, some of the drivers will refuse. DeMille, unscrupulous and inspired as ever, will get his fifteen-year-old daughter Cecilia—an accomplished rider—to perform the feat, shaming the grumbling cavalrymen into agreeing to his demands.

PARTING THE RED SEA

For this, the stand-out sequence of the movie, you will assemble on SEAL BEACH, about 40 feet from the water, positioned between posts and wires set up along the shoreline to create a channel for you to run through. The atmosphere on set will be tense due to the logistical challenge facing DeMille: shooting must begin at midday exactly, because once the sun passes the meridian, the shot will be ruined by the shadows cast across the beach.

At 11:45am, everyone is ready. The Palm Court Orchestra will play "Largo" from Dvorák's *New World Symphony* to establish a mood, despite the sand playing havoc with the brass section. But then DeMille, standing on a raised platform, will suddenly notice that the shore looks much more pristine than the bottom of the Red Sea would. With the clock ticking, the director will look desperately out at the ocean, only to see that there are considerable quantities of kelp floating on the surface. He will leap down and, calling out to you all to join him, plunge into the sea, gathering as much SEAWEED as he can carry and start scattering it on the sand. Hundreds of you will follow suit. Within ten minutes the shore resembles a sea floor and, at noon precisely, DeMille will blow a whistle and you will race through the channel, completing the shot just in time.

Towards the end of the afternoon, you will huddle with your fellow Israelites by the waves as Moses, standing on a rock, assures you that Pharaoh's army will be consumed by the sea. As you pretend to watch the chariots approaching, you will be expected to express a range of emotions: at first doubt and uncertainty, then fear, then awe, excitement, and relief in quick succession. However, just as the cameras are about to roll and the sun begins to set over the Pacific, clouds will obscure its rays, endangering the whole shot. But then, miraculously, the clouds will part and the sun reappear, a magic moment that produces a reaction from you that DeMille could only have dreamed of. As Rita Kissin, a Jewish writer and fellow extra, would recall: "a gasp went through the crowd. The faces of the men and women reflected the light, tears trembled on wrinkled cheeks, and sobs came from husky throats. For many, the world had moved back 3,000 years."

Note that the illusion of the RED SEA PARTING, and then swallowing, the Pharaoh's army, will be conjured up by the SPECIAL EFFECTS DEPARTMENT, headed by Roy Pomeroy, during post-production in Los Angeles. Two large blocks of gelatine, with waves carved into them, will be set on a huge table, split down the middle, and fitted with gas jets. As the table is cranked apart, the jets will be lit, melting the gelatine and giving the impression of moving walls of water. When the resulting footage is run forward, it resembles the sea parting; run backwards, the sea closing.

MOUNT SINAI

When Moses descends from the mountains with the tablets bearing the TEN COMMANDMENTS, you will be in the anxious and expectant throng waiting to hear his words. Gazing down on you, DeMille is not impressed. As far as he's concerned, you are not bringing the necessary emotional intensity to this climactic scene. Crafty as a fox, he will call a break and consult with his right-hand man, "Hezi" Tate. Five minutes later, he will announce over the PA that one of the actors has suddenly died, leaving a wife and eight children, and then ask for a few moments' silence before resuming shooting.

This trick works. Shocked and bewildered by the news, you will react convincingly to a tongue-lashing from Moses for your idolatry, be horrified when he smashes the Holy Tablets, quake with genuine fear when lightning rents the Golden Calf in two, and run around in the required state of abject panic.

DEPARTURE: TUESDAY, JUNE 26TH

With all the major extras scenes done, you will board the train for your return journey to Los Angeles on the morning of June 26, exhausted but strangely elated. At Union Station a guide will meet you for the time voyage home.

Naturally, you will want to view the end result of the movie and see yourself in the finished film. On your return to the present, *The Ten Commandments* is of course available online or on DVD. But how much better would it be to sit in an enthused, excited

SPOT YOURSELF ON THE BIG SCREEN! FOR A SMALL ADDITIONAL FEE, WAG OFFER TIME TRAVELERS THE OPPORTUNITY TO RETURN TO LOS ANGELES FOR THE MOVIE'S GALA PREMIERE.

audience—including celebrity guests—in a packed cinema to see the first public screening of the film at GRAUMAN'S EGYPTIAN THEATER? For a small additional fee, you can return to Los Angeles for the evening of TUESDAY, DECEMBER 4, 1923 and bask in reflected glory as you watch yourself up there on the silver screen, a small part of movie hitory.

The Birth of Bebop

FEBRUARY 15–16, 1942 ✵ HARLEM, NEW YORK

DURING THE SECOND WORLD WAR, JAZZ music underwent a profound evolution, and a new style was born—bebop. Drenched in the blues, this fast, rebellious music, with its insider cool and outsider status, had an incalculable influence not only on jazz, African-American culture, and popular music, but also on literature (the Beats, especially Kerouac), art (the Abstract Expressionists), comedy (modern stand-up pioneers such as Lenny Bruce and Mort Sahl), European cinema (the French New Wave), and fashion (from movie stars to Mods).

On this forty-eight-hour trip, you will get a ringside seat at legendary venues, experience the humming buzz of New York nightlife, dance till you drop at the Savoy, the world's premier ballroom, and attend the ultimate jazz laboratory—the jam session—where Young Turks and giants of the swing era share the same stage together, interrogating and reinterpreting jazz standards. Out of this dialogue between the generations, bebop is born. The trip is designed so you see and hear this collaborative process in action; you will enjoy both the glorious power of the Big Bands and the jagged dynamism of the New Thing.

BRIEFING: THE BEBOP REVOLUTION

This trip is all the more remarkable because the genesis of bop was lost to posterity. A strike over royalty payments called by the American Federation of Musicians, which lasted from July 31, 1942, to November 1944, banned all union members (and if you wanted to work you had to be in the union!) from doing *any* sessions—whether in the studio, on radio, or live—for *any* recording company. As a result, when artists like Charlie Parker and Dizzy Gillespie unveiled their music at the end of the war, it would seem like it had come from nowhere, fully formed, a perception that caused a damaging rift within the jazz community between modernists and traditionalists and obscured bebop's true roots. Thus, you have the unique opportunity to visit a moment in time when they all played on the same side of the street.

You'll be spending time at three key venues in New York, all of which are based in the city's jazz heartland of Harlem.

THE SAVOY BALLROOM

Opened in 1926, and taking up an entire block on 140th Street and 596 Lenox Avenue, the Savoy—*The Home of Happy Feet*—is the leading jazz dance venue not only in New York but the whole world. Dance crazes invented here soon spread across the USA and onto Europe. Owned by the entrepreneur Moe Gale and run by Harlem businessman Charles Buchanan, it costs half a million dollars a year to run, while taking twice that at the door. It is also home to BATTLES OF THE BANDS or CUTTING CONTESTS, during which the house band takes on out-of-towners.

For much of the 1930s the CHICK WEBB ORCHESTRA ruled the roost at the Savoy. The diminutive drummer and bandleader (severe spinal deformities led to his death in 1939 aged just thirty) was a genius behind the kit and the first to record that perennial favourite "Stompin' at the Savoy." His all-conquering outfit, featuring the young Ella Fitzgerald, outplayed Benny Goodman in 1937 and Count Basie a year later.

For your visit, the Savoy's house band will be the LUCKY MILLIENDER ORCHESTRA, a hard-swinging outfit with a penchant for R&B-inflected monster grooves.

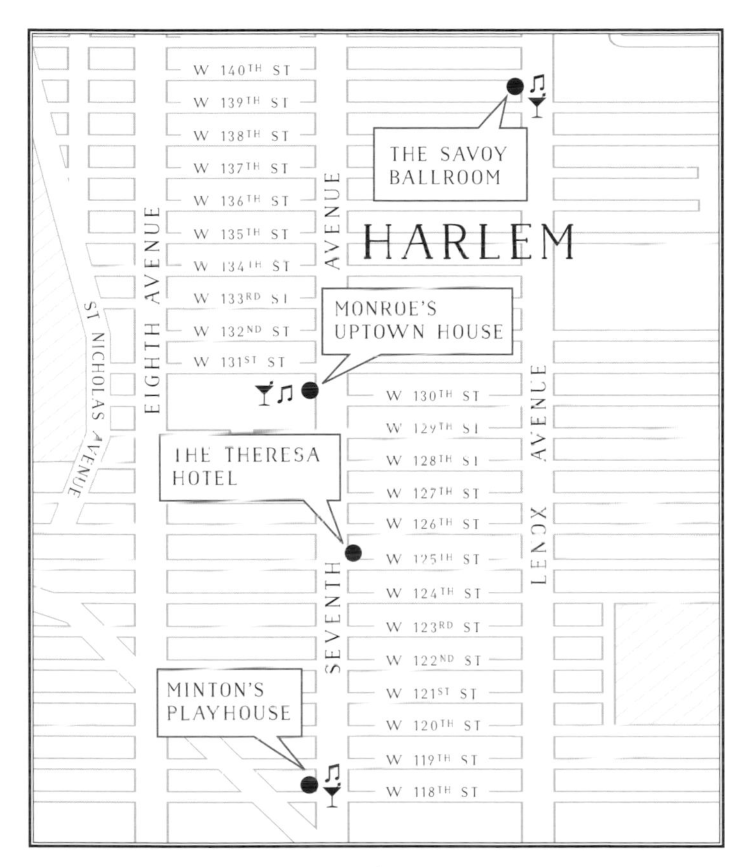

MINTON'S PLAYHOUSE

Minton's is the focal point of the New York jamming scene and is located in the first-floor dining room of the Hotel Cecil, an elegant five-storey building on West 118th Street, near Seventh Avenue. It's owned by HENRY MINTON, a former tenor-sax man, and the first-ever black union delegate, who had previously managed the Rhythm Club—which hosted big names like Louis Armstrong and Fats Waller. With the musical tides flowing in fresh directions, Henry closed and then reopened Minton's in 1940 and hired the ex-bandleader TEDDY HILL to run it.

Hill immediately installed a new house band brimming with young talent. Most notably, he recruited as his musical director the explosive drummer KENNY "KLOOK-MOP" CLARKE, whose off-beat blows on the snare and bass drums had already raised eyebrows in the bands he'd played with. His redefinition of the drummer's role—less tied to holding the beat and interacting more with the soloists—was one of the foundations of bebop. Clarke then brought in THELONIOUS MONK, a mercurial genius with a highly individualistic approach to the piano who was already on the verge of composing masterpieces like " 'Round Midnight" and "Ruby, My Dear."

But what drew swing-era giants like Benny Goodman, Count Basie, Coleman Hawkins, and Lester "Prez" Young to Minton's was the presence of CHARLIE CHRISTIAN on guitar: the most precocious axeman of his generation, he'd cut his teeth in the brilliant small groups of Basie and Goodman, and pioneered a new approach to his instrument, which anticipated rock 'n' roll—though, sadly, this hard-partying cat was hospitalized with TB in 1941, and passed away not long after.

MONROE'S UPTOWN HOUSE

Known as THE PIRATE'S DEN, Monroe's is the go-to place for an after-hours jam, and is located at 198 West 134th Street in a dingy cellar-like basement—appropriately enough, given its underground reputation. It is run by CLARK MONROE, a dedicated jazzhead who inherited the club from the gangster Barron Wilkins after he was gunned down by a pimp called Charleston.

THE TRIP

You will arrive at 3:45pm on Sunday, February 15, 1942, on the CORNER OF 125TH STREET AND SEVENTH AVENUE amidst the crowds of Harlemites gathered for their Sunday promenade, strutting their stuff and dressed to impress. For the duration, you will be based on Seventh Avenue, the broadest boulevard in HARLEM, covering twenty-three blocks and referred to as *Black Broadway* or *The Great Black Way*. Most of the places you visit will be along here or within walking distance on Lenox Avenue, which runs parallel to Seventh and almost matches it for length (twenty-one blocks) and reputation: Langston Hughes, that fine Harlem Renaissance writer, dubbed it *The Heartbeat of Harlem*, and it inspired his poem "Lenox Avenue: Midnight on the legendary street."

To overcome the COLOR BAR that runs through American society, and to ensure that you can enjoy your trip (whatever your ethnicity) without encountering any hassles, you will assume the identity of a French aristocrat. As a fugitive from Nazi Europe, you have been traveling the continent before reaching the Big Apple. Your English is passable but not fluent. You love jazz.

Ever since that amazing performer Josephine Baker (*The Black Pearl*) hypnotised Paris in 1925, jazzmen have been big stars in France; revered as great artists and, most importantly, treated like human beings. For decades to come, the city will offer them a place of refuge. As a result, the tight-knit jazz community will welcome you with open arms. At the same time, being a member of the European aristocracy will ease your entrée into New York's highlife (the VIP PACKAGE).

SUNDAY, FEBRUARY 15TH

Right in front of your arrival point is the HOTEL THERESA, the most upscale, exclusive hotel in Harlem—a very tall white building with Gothic features that takes up a whole block. Owned by Walter Scott, it is a magnet for the big names in entertainment—you might meet Duke Ellington in the elevator—plus business tycoons and politicians. Permanent residents, paying $8 a week to rent a room, include NAT KING COLE, the bandleader JIMMIE LUNCEFORD, and pop sensations THE INK SPOTS.

As you approach the marquee entrance, you will notice taxis and limos disgorging guests, many from Chicago and Detroit, carrying sets of golf clubs—the ultimate status symbol—and wearing riding gear—the latest fashion for those keen to flaunt their wealth. You will go through the double doors into the wide foyer and lighted lobby with its green-and-beige floral wallpaper, and stroll past glistening mirrors towards the registration desk, manned by the slender, golden-brown Eloise Scott. Seated just behind her is Miss Mattye Jean, the college-educated switchboard operator. Adjacent to them are the elevators and three public phone booths.

Your $2-a-night room is on the first floor. Wandering the corridors, decorated by white square tiles with black stripes, you may smell the unmistakably pungent aroma of MARIJUANA, which is referred to as *Reefer*, *Muggles*, *Mary Jane*, or simply *Tea*. It is a staple of the jazz life and there exists a whole subgenre (*viper music*) of songs celebrating the mighty herb: "Gimme a Reefer," "Reefer Man," "Reefer Hound Blues," "The Stuff Is Here and It's Mellow." If you don't come across any at the Theresa, you definitely will at Minton's and Monroe's. Rather than trying to score any dope yourself, which could prove hazardous, you are advised to loiter outside the clubs or out back by the kitchen exits, where much crafty toking will be taking place. Play your cards right and you will be able to drag deep on the odd joint.

By the deluxe double bed in your comfortable room, you will find a copy of *Cab Calloway's Cat-o-logue: A Hepster's Dictionary*. The veteran bandleader (his biggest hit is "Minnie the Moocher") has put together a collection containing 200 *jive* (SLANG) terms, the preferred argot of the jazz world. Take a few minutes to absorb some key phrases that might come in handy during your stay: *hip* = wise and sophisticated; *blow their wigs* = show great excitement; *dime note* = $10 bill; *early black* = evening; *fine dinner* = good-looking woman; *slide your jib* = talk freely.

In the wardrobe you will find three different OUTFITS. For tonight, men will wear a tuxedo with necktie, while women will slip into a sequinned evening gown and white gloves. For daytime tomorrow, both genders will be dressed more conservatively: men in a dark-blue suit, women in a plain dress. For your jamming evening, men will sport zoot suits—broad shoulders, wide lapels, pinstripes, and turned-up baggy trousers—and women a brightly patterned top and a pencil skirt. It's February in New York, so men will be provided with a trench coat, women with a floor-length mink (fake).

At 4:30pm exactly, switch on the RADIO in your room; it will be pre-tuned to a local station that broadcasts a fifteen-minute performance by the new arrivals at the Savoy Ballroom. This Sunday afternoon it's the JAY MCSHANN BAND, a hard-swinging outfit from Kansas City who have on alto sax the twenty-one-year-old

THE BIRD TAKES FLIGHT. CHARLIE PARKER (FAR RIGHT), PLAYING IN THE JAY MCSHANN BAND.

CHARLIE PARKER (*Yardbird* or simply *Bird*). This is only Bird's second visit to the Big Apple. His first saw him gain a few admirers but no paid work; he had to stay with a friend and wash dishes in a club for small change. Pretty much an unknown quantity up to this moment, the broadcast you are about to hear will turn the spotlight on him.

The Jay McShann Band will do five numbers; Bird will be the featured soloist on "Cherokee," a fiendish tune with sixty-four bars and tricky harmonies. Bird's improvisation—juggling the chord progressions, dancing both behind and in front of the beat, waiting patiently for exactly the right moment to unfurl blistering runs of notes—stuns everyone who hears it. Ears prick up all across New York, the five boroughs and beyond. His debut radio performance will draw packs of hungry jazzheads to the Savoy tonight to sample Bird in the flesh—and you will be there too.

STOMPIN' AT THE SAVOY

When you are ready, leave the Theresa and walk the few blocks to the SAVOY, where queues to get in will already be forming; the cost of admission before 8pm is 60 cents, after that, 85 cents. Stepping into its spacious lobby you will be struck by its ostentatious glamour; more like a palace than a dance hall. Hanging above you is a massive cut-glass chandelier; ahead of you, a curved marble staircase. There is a roomy check-in for your coats, and carpeted and mirrored lounges to *take five* (try not to say "chill out"). Climb the staircase and enter the huge ballroom—10,000 square feet, with 4,000 capacity. This giant space is divided into two: one zone has tables and settees where you can sit, watch the action, rest weary feet, and drink cheap beer; the other is for dancing.

The dense, heaving crowd includes all types of people of all ages. Such is the Savoy's fame, you might glimpse a film star or two; Clark Gable and Greta Garbo have both been spotted here. You will notice a dozen well-built, heavy-set men in tuxedos prowling the room. They are the bouncers—ex-boxers and basketball players on $100 a night—under the command of JACK LA RUE. Troublemakers will be discreetly, but forcibly, removed. Also working the floor are the Savoy Hostesses, *café au lait* girls, mostly from Sugar Hill, a prosperous black neighborhood. Their main job is to make sure everybody is having a good time. Male travelers should avoid getting too friendly, as the Hostesses can be fired for consorting with the patrons. However, if you buy a 25-cent ticket, one of them will be your dance partner for the evening.

The specially sprung DANCE FLOOR, all polished maple wood and lined with colored lights, is a thing of beauty. The main section is known as THE TRACK; 250 feet long and fifty feet wide, it can hold thousands doing much-loved and -imitated dances that originated on this very strip: the *Jitterbug*, *Suzie-Q*, *Hucklebuck*, *Camel Walk*, *Snake Hips*, and the *Flying Charleston*, to name a few. The Track will be dominated by young African-

Americans out to dazzle with their moves, demonstrate their prowess, and look good with it: a local marijuana dealer, MALCOLM LITTLE (street name *Detroit Red*), a regular at the Savoy who would later change his name to Malcolm X, never forgot "the black girls in way-out silk and satin dresses, their hair done in all kinds of styles, the men sharp in their zoot suits and crazy conks." For those who prefer a gentler pace, there is a space reserved for the *Tango*, *Fox Trot*, and other classics; the top dancer here is known as *the Sheik*.

The most ambitious hoofers hoping to prevail in the Savoy's hypercompetitive DANCEOFFS will congregate in the CAT'S CORNER, a ten-foot-square area to the right of the two bandstands, as will a throng of spectators, some taking side-bets on the outcome. The winners will be invited to join the 400 CLUB, allowing them to rehearse in daylight hours alongside the bands. Longevity and stylistic innovation are what counts, and will separate the genuine contenders from the no-hopers. Don't even think about stepping into the arena unless you are really hot to trot.

Also featured in the Cat's Corner are the SAVOY LINDY-HOPPERS (from the dance the *Lindy-Hop*, conceived in 1927 to honour Charles Lindbergh's maiden flight across the Atlantic), professional dancers under the control of their mentor and erstwhile agent, HERBERT "MAC" WHITE, an ex-boxer who came through the bouncer's ranks and has been in charge of the Lindy-Hoppers since 1938, recruiting fresh talent and helping to get his dancers work on movies like *Hellzapoppin'* (1941). With a distinctive white streak running through his black hair, he will be hard to miss. Look out for his star performers: FRANKIE "MUSCLEHEAD" MANNING, who has a highly original technique, positioning his body at an acute angle to the floor, like a runner waiting in the blocks for the starting gun, bent low so he can execute his flying aerial leaps; the youngster AL MINNS, known as *Crazy Legs* because of his dazzling speed; and the indefatigable multi-contest-winner NORMA MILLER.

LINDY HOPPERS SWING-DANCING IN THE CAT'S CORNER OF THE SAVOY.

The pumping heart of it all, which keeps everything moving, is the turbo-charged music coming from the bandstands. First up is the JAY MCSHANN BAND. You will see Dizzy Gillespie—with his distinctive horn-rimmed glasses, beret, and goatee, trumpet case in hand—and the inscrutable Monk amongst the coterie of musicians assembled close to the stage, all eager to catch Bird in flight. During Jay's set, BIRD, a slight figure in a shabby, crumpled suit and sunglasses, will solo on "Clap Hands Here Comes Charlie" and "Cherokee," his rapid fingers unleashing sizzling pyrotechnics, driving the band to ever-heavier heights.

Having left the bandstand steaming, it's the turn of the resident outfit, the LUCKY MILLINDER ORCHESTRA, with the wonderful

CLYDE HART on piano (in a few weeks Dizzy will join them, having been fired by Cab Calloway because of his onstage antics). Though they are no slouches, Lucky Millinder's crew are outgunned by McShann and co.; once the BATTLE OF THE BANDS commences, Jay's mob will blow them away.

AFTER-HOURS

For POST-SAVOY EATS AND DRINKS, you could do worse than BRADDOCK'S BAR AND GRILL, right next to the famous Apollo Theater, situated on Seventh Avenue and 125th Street; its two-for-one policy is popular with musicians, and the steaks are pretty good. CLUB BARON is a very sophisticated and luxurious venue on 132nd and 437 Lenox; here you will be entertained by the MC Larry Steele, some up-and-coming comedians, and the talented singer ETHEL WATERS. If you fancy something a little less swish, head over to the DAN WALLI CHILI HOUSE on Seventh Avenue between 139th and 140th, often host to impromptu jam sessions.

Otherwise, you can return to the Theresa, which is still very much alive and kickin." You can get a decent meal in the restaurant and then head to the bar, which is open until 2am. The manager, John Thomas, has a team of seven bartenders working under him, all male and wearing white tunics, and you can get a shot of whiskey for 75 cents. Aside from other late-night drinkers, high-rollers, and city gents, you will notice a number of stunningly beautiful women loitering here, some prostitutes, others local girls hoping to land a rich husband, and a smattering of fellow European aristos. If you are not yet ready for bed after the bar closes, then head down to the lobby, where there will still be fun to be had; you may run into members of the Lucky Millinder Orchestra who are staying at the hotel.

For a more rough-and-ready, but no less jazzy scene, check out the Woodside Hotel. Many of the clientele reflect the criminal demimonde inhabited by its owner, Love B. Woods, a slum landlord with flophouses and shabby apartments all over Harlem; you will encounter pimps, gamblers, hustlers, and sassy dames. However, mixed in are the jazz fraternity: the hotel is Count Basie's base in New York (he penned "Jumpin' at the Woodside" in 1939, and Jay McShann's boys are in the house. Head for the basement restaurant, where you will catch the musicians working on their *chops* (technique).

VIP PACKAGE: THE STORK CLUB

For a small additional fee, you will have the chance to check out New York's most exclusive late-night venue, The Stork Club on 3 East 53rd Street. Opened in 1929, the Stork Club is owned by former bootlegger Sherman Billingsley and attracts A-listers like Charlie Chaplin, Bing Crosby, and Frank Sinatra; literary heavyweights like Ernest Hemingway; real heavyweights like Joe Louis (world champion from 1937 to 1949); and major players in business and finance. Throw in presidents and the occasional royalty, and you have the full array of the East Coast elite mingling here.

A limo will pick you up outside the Savoy and whisk you across town to the club, which opens at 11am and shuts at 4am. From the street, the entrance appears fairly nondescript; an ordinary door with a narrow marquee leading to the sidewalk. The only hint of what lies inside is the 14-carat solid-gold chain barring your way. But don't worry; the top-hatted doorman is expecting you. Once inside, be on your best behaviour. Sherman will not hesitate to eject you if you get out of hand. Even big celebs are not immune: Humphrey Bogart is one high-profile victim of Sherman's no-nonsense approach.

The Stork Club provides a variety of spaces to meet your entertainment needs; there is the tropically themed Island Bar; a cocktail lounge; the glassed-in main dining room with a band and dancing; the Cub Room, a small, intimate venue; while upstairs, you will find a Loner's Room for singletons and the Blessed Event Room for catered parties. By now you will have built up quite an appetite, so make your way to the dining room, where the menu offers haute cuisine with an American twist: dishes like *Tournedos Baltimore* (tenderloin steak with mushrooms, onions, sliced veal kidney, and stewed tomatoes), *Pheasant*

Casserole Derby, *Royal Squab Knickerbocker*, and *Aiguillette of Duckling Florida*, followed by Gallic cheeses and rich desserts.

Though you may prefer wine with your meal, you'd be a fool not to try the club COCKTAILS, which are without doubt the best in New York. The barman, JULIUS CORSANI, will mix you up his own creation, the *Julius Special*, made with Cointreau, Jamaican rum, and a twist of lime, or the house speciality, the *Stork Club Cooler*—gin, sugar, the juice of half an orange, shaken, strained, and served with shaved ice. Other unique concoctions include the *Champagne Cocktail Gloria Swanson*—a pint of very dry champagne, cognac, and lemon peel; the *Alexander the Great*—vodka, crème de cacao, coffee liqueur, and fresh cream, shaken together until ice-cold; the *Snow White*—Southern Comfort, vodka, and orange juice; or the *Millionaire Cocktail*—sloe gin, apricot brandy, Jamaican rum, and a dash of grenadine.

At MIDNIGHT, female travelers should gather in the main dining room to participate in a Sunday night ritual. As the clock strikes twelve, masses of balloons will be released from the ceiling, New Year's Eve–style, each one containing a RAFFLE TICKET. Only women guests are allowed to take part in this light-hearted scramble to grab a balloon and be part of the prize draw; you could win anything from a charm bracelet to a motor car (at least three of the balloons have a $100 bill inside).

At closing time, your limo will be waiting to drive you back to the Theresa for a well-earned rest.

MONDAY, FEBRUARY 16TH

Given the late hour you hit the sack, you may opt for a lie-in on Monday morning. However, if you are determined to cram in some sights, the familiar Manhattan landmarks await you: TIMES SQUARE, FIFTH AVENUE, CENTRAL PARK, etc. For lunch, we recommend the RAINBOW GRILL on the sixty-fifth floor of the ROCKEFELLER CENTER, with its stunning views of midtown and the ice-skating rink down below. Opened in 1934, the Rainbow

Grill's menu contains mostly French dishes, prepared to a high standard, while more regulation American fare is available from the cold buffet and hot grill selections.

If you can't summon the energy to venture out of Harlem, then a great spot for lunch is WELL'S CHICKEN SHACK, serving more-ish waffles and fried chicken. Plus, it's a favourite hangout for musicians; you may be treated to a spontaneous solo set from the exquisite pianist TEDDY WILSON—Billie Holiday rose to prominence in his band—who likes *woodsheddin'* (practising) here in the afternoon.

Having eaten, take the time to explore Seventh Avenue. You will see RADICAL SPEAKERS on street corners preaching revolution, real preachers preaching the Word of God, and all manner of folks going about their daily business. Of particular interest are the BLYDEN BOOKSTORE, run by Dr Willis Huggins, and the NATIONAL BOOKSTORE, owned by Lewis H. Michaux; both stock unrivalled selections of African-American literature, history, and political thought.

CELEBRITY NIGHT AT MINTON'S

Having returned to the Theresa and changed into your jazzy evening wear, head over to the Cecil Hotel, with its dark-blue awning extending over the sidewalk. Go up to the first floor, where you will find MINTON'S itself. Walking in you will see a long bar to your left, its stools occupied by big-wig politicos, penniless dancers, wannabe gangsters, and anxious aspiring jazzers hoping to sit in during the jam and leave with their reputations intact. The barman, Herman Pritchard, dispenses beer at 25 cents a glass and fine sippin' whiskey at 35 cents a pop. To your right is an extremely attractive coat-check girl, who will look after your coat and *lid* (hat). Passing through swing doors, you will enter the music room; on the left-hand side is the tiny bandstand about twelve inches off the ground, and an equally small dance floor in front of it. Behind another set of doors is the kitchen. Take a pew at one of the dozen tables, each one draped in a white tablecloth and displaying flowers in a little glass vase, and tuck into Minton's excellent home cookin': fried chicken, ham hocks, barbecued ribs, grits, black-eyed peas, and hot biscuits.

Tonight is CELEBRITY NIGHT, sponsored by Bobby Schiffman, manager of the Apollo Theater, which is closed Mondays. Minton's offers an open invitation to the bands booked in there, plus free drinks and munchies. This, and Minton's reputation as a happening place, guarantees their presence. By 10pm the room will be packed for the beginning of THE JAM. On stage already are KENNY CLARKE, the prodigal nineteen-year-old pianist BUD POWELL, who is *depping* for Monk (though Thelonious will probably sit in on a few numbers), and ebullient tenor man DON BYAS. Soon they will be joined by the likes of BIRD, DIZZY, BILLIE HOLIDAY,

THELONIOUS MONK, HOWARD MCGHEE, ROY ELDRIDGE, AND TEDDY HILL OUTSIDE MINTON'S PLAYHOUSE.

COLEMAN HAWKINS, HOT LIPS PAGE, BEN WEBSTER, LESTER YOUNG, and many more. At some points there will be as many as fifteen musicians on the bandstand. Not that it's a place for the faint-hearted. Any jammer who hasn't got his shit together won't survive long. The others will carve him up with high-velocity tempos, unusual chord patterns, and esoteric key signatures.

Nevertheless, aside from the odd wounded apprentice, the atmosphere is collegial, though when the temperature rises you'll see two titans of the tenor, LESTER YOUNG and BEN WEBSTER, locked in combat: barman Herman Pritchard compares their duels to dogs fighting in the road. But it's a communal, collective vibe that prevails as the musicians astound and dazzle the audience and each other with their takes on standards like *Body and Soul*, *I Got Rhythm*, *How High the Moon*, *April in Paris*, and *Get Happy*. It's Dizzy who best sums up the mood: "on Monday nights we had a ball."

JAMMING AT MONROE'S

You may want to stay at Minton's until it closes at 4am, but a number of the younger players—Bird, Diz, et al.—will leave at around 2am to get to MONROE'S for when its jam begins. Be prepared for considerably less salubrious surroundings. The club is essentially a basement dive with a rudimentary bar, small kitchen, and barely adequate stage. Yet it exudes an edgy, dark cool befitting its owner, CLARKE MONROE, a hustler who knows all the pushers and dopers, and fences stolen goods in his spare time—silverware, jewellery, furs, and watches. He is here every night from midnight, digging the sounds and making deals. Handsome, charming, six feet tall, with light-brown skin, straightened hair, flashy suits, glossy shoes, and women hanging offa him, Clarke will stand out.

What transforms his venue from a seedy fleapit into a place of magic is the music. Zinging with ideas after their workout at Minton's, BIRD and co. will join the house band—AL TINNEY, piano; RUSS "POPEYE" GILLON, trumpet; VICTOR COULSON, tenor sax; EBENEZER PAUL, bass; and the teenage MAX ROACH, who will go on to raise jazz drumming to a new level—for hours of experimentation. You will hear them stretch time, rhythm, and harmony to the limit, take snatches of melody and fragments of riffs and run with them, speaking a language that will translate later into bona fide bebop tunes; every note hinting towards the future, every bar leading jazz into the unknown.

Monroe's shuts its doors at 7am. Stumble out into the light of day, head singing, and go grab breakfast at SIMON JOUB'S RESTAURANT on 161st and Lenox, the epicentre of the growing Latino presence in Harlem. Not only can you get strong black coffee and yummy sweet and savoury Cuban *pastelitos* (pastries)—made from puff pastry and filled with either cream cheese, guava, pineapple, and coconut, or ground beef in a tomato sauce with raisins and green olives—you will also be nodding your head and tapping your feet to some seriously infectious beats; the set of conga drums in the corner is always in use.

DEPARTURE

Once you've had your fill at Simon Joub's, make your way back to your room at the THERESA, from where you will DEPART.

The Beatles in Hamburg

1960–1962 ❋ HAMBURG, GERMANY

EACH YEAR, HALF A MILLION VISITORS from across the world—many not even born when the Fab Four were making records—make the pilgrimage to a nondescript zebra crossing next to Abbey Road studios in northwest London to recreate the band's celebrated 1969 album cover. But how much more fun to hang out with the Beatles themselves—and before they were famous. This trip offers the chance to follow the original Fab Five as they transform themselves from a teenage beat group into a successful working band during a series of engagements in Hamburg. This will lead them to the cusp of Beatlemania and enduring fame and fortune. To chart their progress, you will take THREE SEPARATE TRIPS, each in a different year—1960, 1961, and 1962—to see three different gigs at three different venues on three separate weekends.

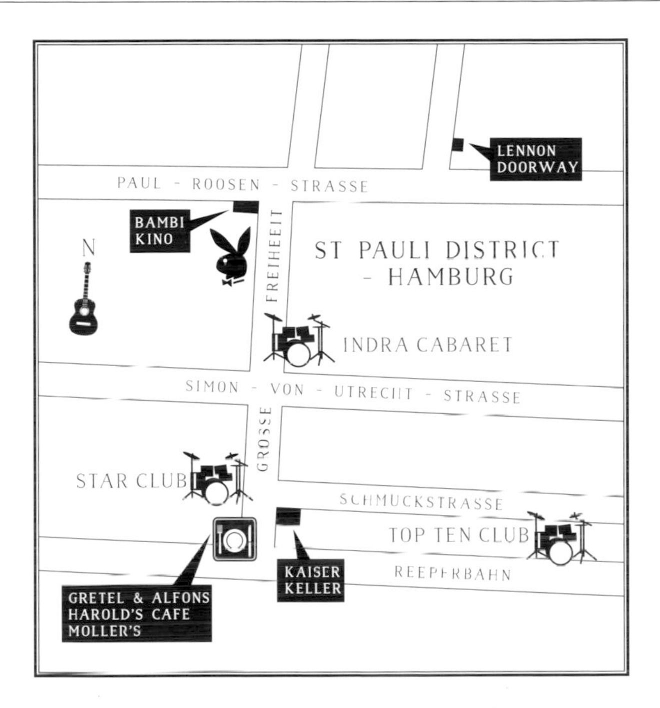

BRIEFING: ST PAULI

Each visit is designed to reflect the Beatles' own experience of Hamburg; the money in your pocket, the clothes on your back, the hours you keep, the places you hang out, will all be dictated by the Beatles' own habits and lifestyle at the time. This way, you will be able to chart the changes they went through en route to stardom.

You will spend the majority of your time in St Pauli, a tough working-class neighborhood close to the docks and home to the notorious red-light district. Nowadays, the area is something of a tourist attraction. At the time of your visit, it is a quarter avoided by most Hamburg residents, due to its reputation for violence and crime. St Pauli's transient population of prostitutes, pimps, and assorted low-life, its sex shops, strip clubs, bars, and brothels, make it the favoured haunt of the Hamburg underworld, and attract tough guys of all shapes and sizes. Adding to this volatile mix is a constant stream of sailors and seamen looking for a good time while on shore leave. As a result, fights are constantly breaking out and murders are not uncommon. It is in this heated environment that the Beatles become the Beatles.

TRIP ONE

On Monday, August 8, 1960, Bruno Koschmider, a squat, hard-nosed ex–circus performer and owner of the Kaiserkeller club in Hamburg, is in London trying to find a five-piece band to open his new club, the INDRA CABARET, in a week's time. Allan Williams, the Beatles' booking agent, convinces Bruno to hire them and they are contracted to play nightly from August 17th to October 16th, with Mondays off.

After a scramble to sort out their passports—none of them had ever left the country before—JOHN, PAUL, GEORGE, STUART SUTCLIFFE (Stu), and PETE BEST (who was reluctantly added at the last minute, making him the group's fourth drummer in thirteen weeks), make the thirty-six-hour journey, Williams driving them in his Austin J4 minibus to save money. They travel down from Liverpool, through London (the boys' first glimpse of the capital), then on to Harwich for the night boat to the Hook of Holland, followed by a long haul across Germany, arriving in Hamburg just after midnight on Wednesday, August 17th. That evening, they step out on the Indra stage for the first time.

FRIDAY, AUGUST 26TH, 1960

You will arrive nine days later, at 5pm, in the darkened stalls of the BAMBIKINO, a pornographic movie theater. It is around an hour into the afternoon feature. A few feet away from you, the Beatles will be just stirring from their slumbers in their tiny digs in the back room of the cinema, located beyond the screen with its own side entrance.

While you may be tempted to remain in your seats to catch the climax of the film, we suggest you slip out early and explore the area while it's still light, and less dangerous. However, before leaving the BambiKino, check out the toilets, as this may be your first chance to cast your eyes on the band. This is where they wash,

THE FAB FIVE IN THEIR LEATHERS, PHOTOGRAPHED BY ASTRID KIRCHHERR: FROM LEFT, PETE BEST, GEORGE, JOHN, PAUL, AND STU SUTCLIFFE.

shave, and clean up before their night's work begins. Stepping out onto the street, you will find yourself directly opposite the INDRA club at 64 Grosse Freiheit, with its crimson-red frontage, Indian elephant signage, and gated entrance adorned by a guitar and saxophone.

Male travelers will be DRESSED as the band: Teddy-boy style, with quiffs, shades, lilac jackets, black shirts, black drainpipes, and crocodile shoes; women travelers will be wearing cream or polka-dot dresses with the skirt above the knee, bobby socks, and flat shoes. You will each have 30 deutschmarks SPENDING MONEY, the same amount as the band gets paid per night. This is a comparatively meagre sum, less than the wages of a manual labourer. Therefore, you will have to be economical, which may prove tricky, as there will be no shortage of people trying to con you out of your last *pfennig*. As the Beatles have no fixed abode, neither do you, but you don't need one, as you (like the band) will be pulling an all-nighter.

The Beatles spend most of their time on GROSSE FREIHEIT, and so will you. Note the contrast between the tall, older buildings, some still bomb-damaged from the war, and the squat, almost shack-like ones hosting an array of bars, sex clubs, and fast-food joints, patrolled by hulking employees "barking" for your custom, and watched over by a hoarding of a giant bikini-clad lady, a fake Eiffel Tower on the horizon—the whole thing is reminiscent of a Wild West or Gold Rush town. As you walk the narrow cobbled street, you will see a substantial venue next to a strip joint, Studio X; this is the KAISERKELLER, where the band will have an ill-fated residency later in their stay.

Reaching the end of Grosse Freiheit, turn left onto the main drag, the REEPERBAHN (Ropemaker's Way). Somewhat glitzier, busier, and a shade more cosmopolitan, this offers the same range of facilities but on a slighter grander and more expensive scale. Don't be fooled, though—it's still very rough and ready. Here neon rules the night, turning the whole strip into a flashing and blinking carnival of sex and debauchery.

Those keen to see more of St Pauli's seamy underbelly should turn off Reeperbahn into DAVIDSTRASSE, which leads to the docks, and take the third right into HERBERTSTRASSE, a dingy alleyway where prostitutes stand in windows offering their wares.

THE GIG

Aim to be back at the Indra by 8pm. The venue is tiny, with the small stage at the back of the room, a heavy, faded, red curtain behind it, and tables gathered in front. The clientele are a mix of strippers, hookers, transvestites—drag acts are popular here and at many of the St Pauli clubs—some sailors, and a scattering of "rock kids," young local music fans who have summoned the courage to venture into Grosse Freiheit at night. You are best off taking a seat as close to them as possible. Chances of a brawl breaking out are fairly high, especially as the waiters, often ex-Nazi thugs with a grudge, are looking for any excuse to start one. According to George Harrison, "all the waiters had tear gas guns, truncheons, knuckle-dusters." As far as drinks go, bottled beer is the cheapest option; avoid the over-priced schnapps and ersatz champagne.

The Beatles will be doing four one-hour sets: 8:30–9:30pm, 10–11pm, 11:30pm–12:30am, and 1–2am. Filling this amount of time will stretch the band's catalogue of covers to the limits; you will be treated to versions of the whole of their favourite albums by artists like Carl Perkins, Elvis, and Gene Vincent, some Jazz Age standards like "Summertime," "Somewhere Over the Rainbow," and "Moonglow," classic hits by Chuck Berry and Fats Domino, current chart-toppers such as the Shadow's "Apache," and a fifteen-minute version of Ray Charles's R&B big hit "What'd I Say."

Their renditions will be often clumsy and incoherent, not helped by the constant flow of free drinks sent to them by appreciative punters, and the fact that Paul is on rhythm guitar, an instrument he is uncomfortable with, Stu is almost a beginner on bass, and Pete is a barely adequate drummer, unable to keep time properly. To try to help him hold on to a beat, the rest of the band will stomp their feet hard on the stage; Pete will respond by giving his bass drum a hefty kick—thomp, thomp—raising the noise level to such an extent that the old woman who lives above the Indra is forced to complain to the manager, Wilhelm "Willi" Limpensal. Despite these flaws, it's hard not to be impressed by the boys' sheer energy and charisma, enough to send a shiver up your spine. Most startling will be their attitude, not dissimilar to early punk bands; you will see them spit, belch, curse, and hurl abuse at the crowd. John is particularly fond of taunting them about Hitler.

At 9:45pm, you will hear an announcement come over the PA, and fifteen minutes later you will notice members of staff going around to check the IDs of anyone young-looking—it is against the law for anybody under the age of eighteen to be in a club after 10pm—and chucking out all the offending punters. Luckily, nobody seems to mind that George is only seventeen!

POST-GIG

After the Indra closes at 3am, you may well be in the mood to hit the Reeperbahn and check out other clubs like the BEER SHOP, MAMBO, WAGABOND, and CHUGS. Don't: they all operate a system known as the *nep*—"additional charges" added to your bill so that when it arrives it will be considerably larger than you expect and considerably more than you can afford. Failure to pay will result in a severe beating.

Instead, head over to HAROLD'S CAFÉ at 15 Grosse Freiheit, one of the Beatles' favoured post-show hangouts. There you can get cheap eats—burgers, fries, hot dogs—and bottled beer. Occupy one of the booths with plain wood tables and hopefully you will see the band come in and order their regular breakfast of cornflakes and milk, a choice that reflects their lack of cash and sophistication.

DEPARTURE

By 8am, the Beatles will be ready to stagger off to their makeshift beds at the BambiKino. This is your cue to leave. Walk along Grosse Freiheit to the corner of Paul-Roosen-Strasse. Turn into it and proceed to Wohiwillstrasse—second on the left. Carry on a few yards until you reach an alleyway that leads into an OPEN COURTYARD flanked by tenement buildings. You will depart from here, the spot where a few weeks later John will have his photo taken standing nonchalantly in a doorway, an image that will later grace the cover of his 1975 *Rock 'n' Roll* album, which features many of the tunes bashed out night after night in Hamburg.

TRIP TWO

The Beatles' stint at the Indra ends prematurely on October 4, 1960, due to the din they are making. Bruno Koschmider shifts them over to his Kaiserkeller, but there things go from bad to worse. Their already fraught relationship with Bruno descends into open conflict. Matters come to a head after George is deported when the authorities find out his true age, while Bruno has the rest of the band arrested for damaging the stage. By December 10, 1960, John, Paul, and Pete have all been sent packing.

One bright spot during this period is their burgeoning friendship with two young Hamburgers: Klaus Voormann, an art-school graduate, and Astrid Kirchherr, a graduate in fashion and photography. At twenty-two, they are a touch older than the band and will open their eyes to different aspects of the city beyond St Pauli. Stu and Astrid quickly fall head over heels in love, getting engaged on November 28th, and Stu remains in Hamburg, where he plans to continue his art studies.

Despite the debacle at the Kaiserkeller, the band had made an impression and are invited back to Hamburg to do a seven-night residency at the TOP TEN CLUB. On Tuesday, March 28, 1961, John and George, traveling by train this time, set off from Liverpool Lime Street, arriving via the Harwich–Hook of Holland route at Hamburg's main station, Hauptbahnhof, at 3:16am on Thursday, the 30th, where they are met by Stu and Astrid. Two days later, Paul and (for want of a better alternative) Pete join them, and the band open at the Top Ten on Saturday, April 1st. George will describe this residency as "fantastic."

SATURDAY, APRIL 15TH, 1961

At 5pm on Saturday, April 15th, you will arrive at the courtyard in Wohlwillstrasse with 35 deutschmarks in your pocket, slightly more than last time. Men will be dressed to match the Beatles' latest HAMBURG LOOK—black leather jackets, black velvet shirts, black leather trousers, cowboy boots. Women will be wearing black leather jackets, black polo necks, black leather skirts, black tights, and boots. All these items have been sourced for you from the uptown HAMBURGER LEDERMODEN, where the band saw the gear that inspired their new style (the store proved too expensive for them, so they engaged the services of a St Pauli tailor to reproduce its clothes at a fraction of the price).

Again you have no accommodation; you will spend another night on the tiles. This time it will be a good idea to fill your belly before the gig; it's never wise to take amphetamines on an empty stomach. Make your way to SCHMUCKSTRASSE, just off Grosse Freiheit, where transvestite hookers linger in doorways, and at number 9 you will find CHUG-RU, a cheap and cheerful Chinese restaurant where the Beatles often go to eat—they are very fond of the pancakes. Once you are done, head to the Top Ten Club at 136 Reeperbahn.

THE GIG

You will enter the TOP TEN CLUB, housed in an old, narrow-gabled building, with an extended lip over the entrance bearing large blue signage, at 7:30pm. Consisting of one large room, the raised stage is set alongside one wall, the sizable dance floor (*tanzfläche*) directly in front of it, the bar to its right.

You may well encounter the owner, HORST FASCHER, a thirty-six-year-old former champion amateur boxer who did nine months in jail for killing a man in a St Pauli street fight, and you will certainly see his younger brothers, UWE and MANFRED, also boxers, whose job it is to shield the band from any unwarranted attention.

On the whole, however, the Top Ten's clientele is a bit less seedy than at the Indra and there is a larger proportion of rock fans. Horst wants Top Ten to be all about the music and has had his manager, Peter Eckhorn, install a state-of-the-art sound system and Binson Echo microphones—which the Beatles love using.

The Beatles will be playing from 8pm until 4am, with fifteen minutes' break every hour. With Paul on piano, having gladly ditched his rhythm guitar, the band dynamics are better, though some stomping will still be required to keep Pete in line. You also might get to hear Paul on bass, as Stu, now enrolled at Hamburg Art School, is often absent. Belting out a similar set-list to the one they mined at the Indra, you will appreciate how much more together they are, their vocal harmonies smooth and sophisticated.

These improvements are partly due to the fact that they will be joined on stage by TONY SHERIDAN on electro-acoustic Martin guitar. Sheridan is a gifted musician whose promising career as a rock 'n' roller has stalled in the UK because of his chronic unreliability. A Hamburg veteran, he has worked most of the clubs with visiting artists and pick-up bands. The Beatles really dig him—they will go on to record with Sheridan while in Hamburg, their first-ever studio sessions—and are forced to raise their level to match his.

Another factor affecting the band's performance is SPEED—the drug, not the tempo—which they are all taking on a nightly basis, along with pretty much everybody else in the club. Preludin (*prellies*) is the stimulant of choice. Licensed in Germany in 1954, it is an appetite suppressant that is available on prescription from any chemist. Aside from giving them the stamina to keep doing these marathon sets night after night, it also adrenalises them onstage; cranked up, buzzing hard, there is an unhinged wildness to the show, a thrilling edge and danger to every note played.

We suggest you take some Preludin yourself. As a one-off experiment it should do you no harm and will put you in sync with the band and all the other night owls that inhabit the area.

THE BEATLES RAISE THEIR GAME, INSPIRED BY TONY SHERIDAN (FRONT RIGHT), AT THE TOP TEN CLUB.

To get your hands on the pills, simply go downstairs to the Gents, where you will encounter Rosie Hoffmann, the sixty-two-year-old *toiletten frau* who sits at her small table with a bowl for tips and a big glass jar full of what look like mints but are actually *prel-lies*. Rosie, who is well-liked by the band, charges 50 pfennigs a pill. One or two will be quite sufficient for your requirements and will match the band's consumption—except for John, who will be necking a handful over the course of the evening.

At around 1pm you will see a group of well-dressed gangsters swagger into the club, take tables right near the stage, and start lavishing drinks on the band. These are some of the Beatles' most dedicated fans: VIP criminals led by WILFRIED SCHULZ, king of the Hamburg underworld, dubbed "Der Pati von St Pauli" (Godfather of St Pauli) by the press, and including hardened muscle like WALTHER SPRENGER, the proud owner of fifteen convictions for grievous bodily harm. Whatever you do, don't spill your drink on him! The hoods are so into the music that you will see them make requests and even get up on-stage and sing along with some of the numbers.

POST-GIG

Around 4am you will be turfed out on the Reeperbahn. Ignoring the usual temptations, head over to GRETEL & ALFON'S BAR at 29 Grosse Freiheit. Its low ceilings, nautical-themed decor, and cosy fireplace make it feel like an Olde English pub, an impression enhanced by its white exterior, small windows, hanging baskets, and modest signage. The Beatles feel especially at home here; the owner, Horst Janowiak, is letting them all crash in his apartment at 66 Grosse Freiheit. So get the drinks in and enjoy the convivial, intimate atmosphere as you enter the wee small hours.

DEPARTURE

Around nine on Sunday morning, the Beatles are in the habit of strolling over to the BRITISH MARINERS' MISSION at 20 Johannisbollwerk, just a few minutes' walk away via St Pauli Hafenstrasse and adjoining the Gustav Adolf Church. This venerable establishment is open to UK sailors and seamen looking for somewhere to lay their heads—for 4 deutschmarks per night, men only—and offers free fry-ups every morning. The familiar nosh, plus the

British newspapers and authentically strong tea, is what attracts the band. Find yourself a table and tuck into a full English breakfast—liver, bacon, sausage, eggs, grilled tomatoes, mushrooms, and toast/fried bread. After your heavy night, it will really hit the spot and take the edge off your comedown from the Preludin.

Meal finished, leave the Mission. Right next to it is a railway bridge. Go stand in the tunnel underneath it, and you will depart from there.

TRIP THREE

The Beatles' career takes a huge step forward on Wednesday, January 24, 1962, when BRIAN EPSTEIN signs on as their manager. Soon after, they make their first appearance on BBC Radio and start gigging at the Cavern Club in Liverpool. Meanwhile, they are head-hunted by MANFRED WEISSLEDER, who has made a fortune from his chain of St Pauli sex clubs and is planning to open a brand-new rock venue in a converted cinema. Having outbid Peter Eckhorn, who wants the band back at the Top Ten, Weissleder secures the Beatles for a two-week residency, starting April 13th.

This time the money is better and the band gets to fly to Hamburg, but the whole trip is overshadowed by the sudden death of Stu from a brain aneurysm on April 10th, the very day the band lands at Hamburg airport. This tragedy hits them all hard and turns their time at Weissleder's STAR CLUB into something of an emotional endurance test. John is most deeply affected, and his anger and grief leads to onstage craziness. One night he comes on dressed as a cleaning lady, another stripped to the waist with a toilet seat around his neck.

Having survived this nightmare, the band's return to the UK sees them back on an upward curve. On August 16th, having finally fired Pete (Brian Epstein did the honours), Ringo takes on the drum duties. Ringo first met his future bandmates in Hamburg during their ill-fated stint at the Kaiserkeller in 1960 while he was playing with Rory Storm and the Hurricanes, and their paths criss-crossed from then on, both in Liverpool and Hamburg. On September 4th, they complete their first session for EMI, guided by George Martin's cultured ears. On the 22nd they appear on TV for the first time; on October 5th, "Love Me Do" is released; by the end of the month it sits at number 27 in the *NME* charts. However, there is unfinished business in Germany. Epstein has booked them in at the Star Club again from the first to the fifteenth of November, even though the band, as John observed, had "outlived the Hamburg stage and wanted to pack it in."

SUNDAY, NOVEMBER 11, 1962

You will arrive this time in the lobby of the HOTEL GERMANIA, an old building in decent condition with three floors and an attic on Detlev-Bremer-Strasse, about five minutes' walk from Grosse Freiheit. Your room is booked and paid for in advance. You will find it perfectly adequate, with all the basic amenities. The Beatles are also staying here, courtesy of Weissleder's cheque book.

Reflecting the band's new-found celebrity, you will have 73 deutschmarks to burn, over double what you had on the first trip. And there's a new DRESS CODE. Men will be sporting dark-blue mohair suits, white shirts, and slim ties, sartorial choices dictated by Epstein's desire to make the band more appealing to a mainstream pop audience. Women will be wearing black, sleeveless dresses with high necklines and ankle boots. As it is winter and below freezing at night, you will be provided with coats; for the men, knee-length navy-blue duffel coats; for women, dark-grey trench coats.

Looking sharp, you will step out of the hotel and walk south towards Simon-Von-Utrecht-Strasse, which takes you onto Grosse Freiheit. The Star Club is at number 39, next door to an erotic film theater, and instantly recognisable by the large hoarding covered with the names of the big acts who've played there, positioned over the darkened entrance, with the club logo in neon above.

THE GIG

Once inside the Star Club, you will be struck by its size compared to the other venues you have visited: it has a 2,000 capacity and a huge dance floor—a site of exuberant, frenzied, non-stop motion. The stage is standard, with a cityscape backdrop behind the bands. The crowd will be predominantly rock fans, interspersed with the usual suspects. The place will be packed and you may struggle to find a seat until 10pm, when the under-eighteens are all ejected.

Armed with your extra cash, you will be able to afford to drink your fill without fear of retribution. Look out for one of the barmaids in particular—BETTINA DERLIEN. Known as Big Betty, she is a beautiful, buxom, curvaceous, and outspoken character who is infatuated with John—an interest he is happy to repay.

The Beatles will be doing two hour-long sets, one before 10pm, and one late, and sharing the bill with other acts: TONY SHERIDAN; ROY YOUNG, a London-born artist in the mould of Jerry Lee Lewis; DAVY JONES, a Manchester-born singer who will go on to star in America with the Monkees, the first-ever manufactured boy band; and another Liverpool group, KING-SIZE TAYLOR (the six-foot-five lead singer and guitarist) AND THE DOMINOES.

When the Beatles hit the stage, you will immediately notice the difference that Ringo makes to the overall sound of the band by maintaining a rocksteady beat. Gone is the ragged quality of earlier performances. The band will be tight, slick, and in command as they knock out the usual combination of rock 'n' roll hits—"Twist and Shout," "Be-Bop-A-Lula," "Roll Over Beethoven"—selections from the Great American Songbook—"Red Sails in the Sunset," "Falling in Love Again"; and novelty covers like Fats Waller's "Your Feet's Too Big."

IT'S RINGO—AT LAST. THE FAB FOUR SHAKE IT ALL UP AT THE STAR CLUB IN 1962.

The undoubted highlight of the night will be the headline act, none other than the already legendary LITTLE RICHARD and his backing band, SOUNDS INCORPORATED. Born in 1932, this flamboyant, hugely gifted pianist with deep roots in gospel, blues, and R&B, author of such hits as "Tutti Frutti," "Long Tall Sally," "Lucille," and "Good Golly Miss Molly"—all staples of the Beatles' repertoire—is currently at a crossroads in his career as he contemplates abandoning secular music all together in favour of the church. However, this dilemma will not affect his performance, and you will experience Little Richard at his outrageous best. Starting the show in a tuxedo, white shirt, and bowtie, he will slowly strip off his clothes until he is down to just his trousers, which he will remove while standing on the piano to reveal a pair of bulging bathing trunks.

Every night, the Beatles will watch awestruck, learning at the feet of the master, while offstage Little Richard will offer them spiritual guidance. Another bonus for the band will be their friendship with the teenage prodigy BILLY PRESTON, keyboardist with Sounds Inc. Preston will go on to appear with the Beatles on tracks like "Get Back" during their last ever performance, filmed on the rooftops of Apple HQ on January 30, 1969.

DEPARTURE

After the gig, make your way to the junction of Grosse Freiheit and Reeperbahn, where you will find CAFÉ MOLLER. This one-storey, lemon-sherbet building, with a flat roof and a hoarding featuring a large chef in a white hat next to a waitress bearing multicolored ice-cream cones, resembles an American diner. There are tables outside if you want to sit and watch the action on Reeperbahn or, if you'd rather not expose yourself to the cold, take a pew inside and order the Beatles' current breakfast of choice—ham and eggs. Moller's home-made cakes are also highly recommended.

By 8:30am, you should be back at the HOTEL GERMANIA. Given this is the first time you will have access to a bed, you will now have two hours to enjoy this luxury before DEPARTURE from your room.

The Rumble in the Jungle

OCTOBER 29–30, 1974 ✵ KINSHASA, ZAIRE

Sport lends itself to hyperbole, but this fight really was extraordinary. First, the setting: the Rumble in the Jungle took place in the heart of Africa, in a stadium built over a prison of medieval ghastliness in which fifty criminals had just been executed at random to discourage their associates from misbehaving while the eyes of the world were on their country. Then the timing: in an era of elevated black consciousness, the African setting was inspired, while the new technology of satellite television allowed Kinshasa to become the authentic centre of planetary attention, at least for an hour. Next, the cast of characters: the bout was promoted by a vertical-haired ex–numbers racketeer, financed by a dictator who knew no bounds, and featured boxers—Muhammad Ali and George Foreman—who were, respectively, arguably the most charismatic man alive and the Platonic embodiment of brooding menace. Finally, there was the fight itself: it was David vs. Goliath, Age vs. Youth and Beauty vs. Beast rolled into one, a bullfight in which the tables were turned on the matador. There was only one way to

MOBUTU INTRODUCES FOREMAN AND ALI TO ZAIRE, ON THEIR ARRIVAL IN KINSHASA.

beat Foreman, and Ali found it. This was his apotheosis. Even Foreman would ultimately be transformed by the magic of the occasion. You don't often get to see a myth unfolding in real life, but you will tonight.

BRIEFING: STAGING THE FIGHT

The organising genius behind the fight is Don King, a character whose manipulations Machiavelli might have found distasteful. A former illegal-gambling magnate from Cleveland, Ohio, King saw the light, or at least the dollar signs, while serving a four-year stretch for beating a man to death (he owed King $600). He had a dream. It wasn't as nice as Martin Luther King's, but boy it was compelling. He would take over the world of boxing.

On leaving prison in 1971, King managed to entice Muhammad Ali to take part in an exhibition bout in

Cleveland to raise funds for a local hospital. Within three years, he was hatching plans for the most ambitious bout in the history of boxing. Using his sorcerer's tongue, he persuaded both Ali and Foreman to sign contracts to fight one another for the unprecedented sum of $5 million. Now all he needed was someone to put up the cash.

He found what he was looking for in the person of Joseph-Désiré Mobutu, or Mobutu Sese Seko Kuku Ngbendu Wa Za Banga, as the president of the vast central African Republic of Zaire restyled himself in 1972. The name translates as "all-powerful warrior who, because of his endurance and inflexible will to win, goes from conquest to conquest leaving fire in his wake." This has certainly been the experience of his political opponents. Ali probably wishes he had thought of the nickname first.

By systematically creaming off the wealth derived from Zaire's extraordinary natural resources, Mobutu has already become either the third- or the seventh-richest man in the world, depending on whom you speak to. (His subjects are among the poorest, with an average annual income of around $70.) He can thus foot the bill for the fight with ease. And, as King has correctly calculated, staging it deeply appeals to his vanity. Mobutu is no mug, however, and is not about to let himself be outshone by Ali's charisma. Accordingly, he will watch the fight from the considerable comfort of his palace on closed-circuit television.

In future years, King will miraculously appear by the side of almost every new world champion, his grinning visage becoming as much a part of the post-fight ritual as the handing over of the title belt.

PLEASE NOTE: The fight was originally scheduled for September 25th, but has had to be pushed back five weeks due to Foreman sustaining a cut to his eye in training. This scuppered plans for the bout to coincide with the ZAIRE 74 MUSIC FESTIVAL, which took place from September 22nd to 24th at the Stade du 20 Mai. It featured James Brown, B.B. King, Bill Withers, and the Spinners, as well as African artists including Miriam Makeba, Tabu Ley Rochereau, and Franco's TPOK Jazz, and will soon be available as an additional journey for our travelers.

THE TRIP

On the morning of October 29, 1974, you will arrive at the GARE CENTRAL on Kinshasa's PLACE DE L'INDÉPENDENCE. It is less than fifteen years since the Belgian colonists packed their bags: Leopoldville has turned into Kinshasa, and in a short period of relative stability it has grown into a thriving and lushly green capital city, its colonial squares flanked by new hotels, highrises, and the trappings of modern urbanism. You should have no difficulty getting a taxi here.

WEATHER, ROOMS, AND FOOD

The WEATHER will be hot and humid, in the mid-90s F during the day and dropping no lower than 80°F in the middle of the night. The night of the fight itself will be beautiful, clear, and lit by a full moon, but bear in mind that just half an hour after the contest finishes there will be a prodigious rainstorm.

Local CURRENCY is the ZAIRE, and the official exchange rate is currently half a zaire to the dollar. You will be able to do better on the black market. This would be unwise for regular travelers, who on leaving the country have to prove exactly where they have changed money, but as you will be departing in a time machine this shouldn't be a problem. Regular tickets for the fight are going at 5 zaires ($10) a shot, though many locals will find a way to get in for less.

You'll be lucky if you manage to get to bed as early as 7am on the night of the fight (which *begins* at 4am local time), but you'll certainly be looking forward to some sleep thereafter. We can offer rooms at the INTERCONTINENTAL on Ave Batetela, where Foreman and his entourage are staying, as well as the American novelist Norman Mailer. Alternatively, the HOTEL MEMLING, 5 Avenue Rep du Tchad, is a grand old pile

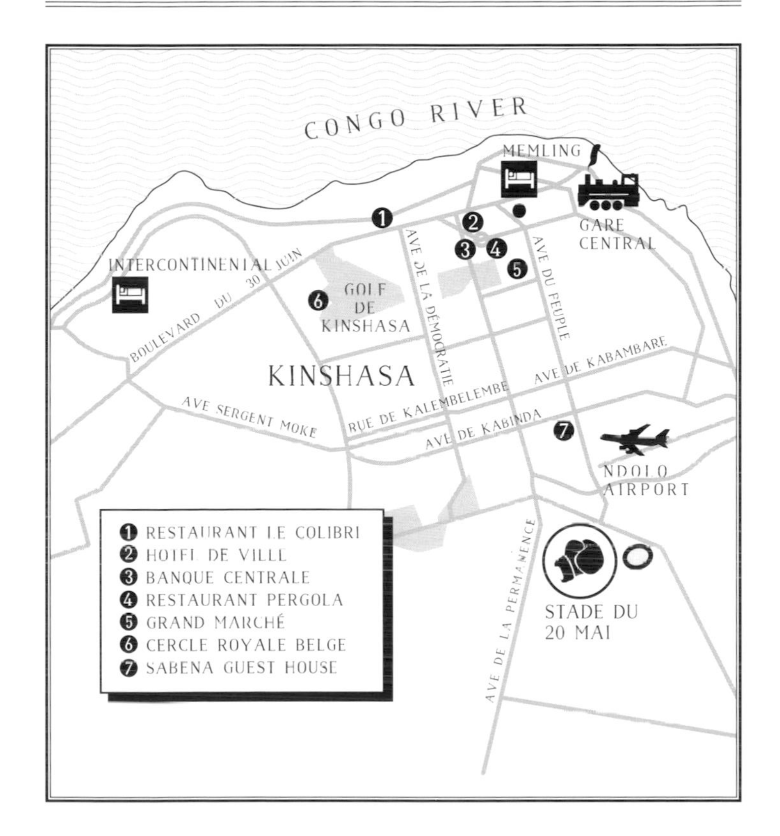

recently refurbished by Belgian airline Sabena. And, close to the airport, the airline has a SABENA GUEST HOUSE, which provides rooms and delightful cottages in its immaculately kept grounds.

For top-end EATING, try the PERGOLA restaurant opposite the BCC (Banque Centrale du Congo) building. Similarly priced is the CERCLE ROYALE BELGE, situated by Kinshasa's golf course (entrance on Avenue du Cercle). Another good choice is LE COLIBRI, a French-style bistro at 61 Avenue Lusaka, founded in the 1950s, whose speciality is *Le Toast Cannibale*—a take on steak tartare.

OH KINSHASA!

Boxing aside, KINSHASA offers travelers plenty of other diversions. Most SHOPPING needs can be met at the fabulous Art Deco SEDEC

BUILDING (Boulevard du 30 Juin), previously a motor showroom and now the city's first self-service grocery store. For African crafts, head for the MARCHÉ DES VOLEURS, where you will find plenty of malachite and ivory trinkets, and Kasai carpets; but do remember our clients are not allowed to bring souvenirs home.

A vibrant post-independence THEATER scene is developing on the Kasa-Vubu Road, but Kinshasa's strong suit is MUSIC, DANCE, AND NIGHTLIFE. In fact, you have arrived in the midst of a golden era and are likely to find superb sounds in almost any neighborhood bar. Among the leading bands of the time playing Kinshasa, look out for ZAÏKO LANGA LANGA, FRÈRES SOKI & L'ORCHESTRE BELLA BELLA, EMPIRE BAKUBA, and, above all, FRANCO & L'ORCHESTRE TPOK JAZZ. You are likely to hear one of these at the SCOTCH CLUB, LE BODEGA (though do bear in mind this is also a house of ill repute), or the landmark LA PERRUCHE BLEU. The HOTEL OKAPI on a hilltop in the Binza Ozone district (west of the old city) has a great dance floor but tends to be very expat heavy.

For a taste of Kinshasa high society, be sure to take in LA DEVINIÈRE in Binza, opposite the presidential Palais de Marbles. Behind its outer walls you will find a delightful garden, fine French cuisine, and leading figures in the government and the Lebanese business community.

THE FIGHT

THE FIGHT will take place at 4am local time in the STADE DU 20 MAI in the Quartier Matonge, about half a mile from Ndolo Airport. Despite the weird hour (to fit with US TV schedules), around 60,000 locals—and a rather poor showing of VIPs—will be attending. The ring sits in the centre of a soccer field, an enormous poster of Mobutu in trademark leopard-skin hat sits behind one of the goals. Most of the locals will be pitching up several hours before the fight between 10 and 11 pm. If you choose to join them, you will be treated to several displays of TRIBAL DANCING.

Take a look at the people in the RINGSIDE SEATS. Alongside many members of Mobutu's entourage, there will be a smattering of American celebrities. Ex–heavyweight champion JOE FRAZIER can be seen in a very groovy green-and-yellow Hawaiian shirt under a maroon sports jacket. Also present are American literary luminaries NORMAN MAILER, GEORGE PLIMPTON, and HUNTER S. THOMPSON. Seats have been reserved for our clients at the back of the VIP zone on the side opposite Mobutu's portrait.

GEORGE FOREMAN

The 1974 version of GEORGE FOREMAN is light years from the cuddly bald electric-grill tycoon of the future. The twenty-five-year-old World Heavyweight Champion is a snarling fighter with a distinct aura of menace. He has got off to a bad start with the people of Zaire by pitching up with his pet German shepherd, a breed indelibly associated with the police force of their Belgian former overlords. He will go into the fight with almost the entire nation rooting for his opponent. But he doesn't seem particularly troubled by this.

As economical with words as Ali is profligate with them, Foreman conserves his energy to maximise his menace. Raised in Houston, Texas, he was by his own admission a teenage hoodlum of the kind you would most definitely not want to meet on a dark night, before finding salvation of a kind in boxing. He stormed his way to Olympic Gold in 1968. Since then he has won all forty of his professional fights, thirty-seven of them by way of a knockout. Two of his recent victims have been JOE FRAZIER and KEN NORTON, both of whom have beaten Ali since his comeback (though Ali has subsequently levelled the scores). Foreman demolished both of them within two rounds, knocking Frazier to the canvas six times.

Foreman is considered one of the hardest punchers in history, as a glance at the heavy bag he uses in training graphically

FOREMAN IN TRAINING. HE PLANS TO DO SOMETHING SIMILAR TO ALI WITH ONE OF HIS "HAYMAKER" PUNCHES.

confirms. After each session, it has an indentation several inches deep. If Ali allows Foreman to corner him, his body will become that bag. At least as frightening as the power of the champion's blows is the relish with which he delivers them. Archie Moore, the wily ex–light heavyweight champion who is co-training Foreman for the fight, is seriously worried he might kill Ali if the latter doesn't do the sensible thing and let himself get knocked out quickly.

During the unforeseen delay, Foreman has holed up in the Intercontinental Hotel in downtown Kinshasa. His daily workout takes him and his crew on a six-mile run along the banks of the Congo river, just north of the hotel.

MUHAMMAD ALI

Where do you begin with MUHAMMAD ALI? Born Cassius Marcellus Clay in Kentucky in 1942, the "Louisville Lip" was one of the most controversial figures of the 1960s and remains a profoundly divisive character going into this fight. He "shook up the world," in his words, by winning the world heavyweight boxing crown at the age of twenty-two with a stunning display against the terrifying Sonny Liston. He hasn't stopped shaking it since.

Shortly after capturing the crown, Ali became a Muslim, changed his name, and joined Malcom X in the Nation of Islam, further alienating the wide section of white America that was already none too fond of this boastful nonconformist. After three years and nine dazzling title defences, the establishment got its opportunity for revenge when Ali refused to be inducted into the US Army to fight in Vietnam ("No Vietcong ever called me nigger"). His boxing licence was revoked and he was not allowed near a ring for the three and a half years that would, under normal circumstances, have coincided with the peak of his boxing prowess. There are those who will never forgive Ali for refusing to fight for his country, but the wheel of history has turned, and his anti-Vietnam stance is now much more in tune with the times.

Ali fights like no other heavyweight boxer. He has the speed of a lightweight schooled in the Shaolin monastery, and can throw a bewildering six jabs per second. He defies convention by holding his hands low, aiming his punches almost exclusively at his opponents' heads, and leaning out of the way of shots aimed at his own, rather than attempting to block them with his arms. As a child, he used to ask his brother to throw rocks at him to sharpen his reflexes, and it shows.

Ali doesn't act like other fighters, either. A promoter's dream, he is an inspired self-publicist who composes poems ridiculing his opponents, predicts the round he will knock them out, and does anything in his considerable power to upset their composure both before and during his fights. Humility is not his middle name—he has been describing himself as "THE GREATEST" for a decade—and his breathtaking self-confidence tends to be self-fulfilling, as well as highly entertaining. But there are many who are itching to see him get his comeuppance.

Ali's performances since his comeback in 1970 have been a mixed bag. Though retaining unfeasible hand speed—he still STINGS LIKE A BEE—the lay-off and his advancing years seem to have compromised his ability to FLOAT LIKE A BUTTERFLY, dancing around the ring and evading his opponents at will. Ali can still undeniably talk the talk, but can he walk the walk? We are about to find out. Ali has based himself in a villa in NSELE, near one of Mobutu's palaces on the eastern side of the city. The gym where both fighters have been training is also located here.

THE TALE OF THE TAPE

The size difference between the two fighters is distorted by their auras and reputations. They are the same HEIGHT (6'3") and, while Ali has a slightly longer REACH, at the weigh-in Foreman is the heavier in WEIGHT (though by a mere 3.5 lbs.). You are prohibited from betting because you know the result, but the most widely quoted ODDS on an Ali victory are 3–1 against.

	MUHAMMAD ALI	GEORGE FOREMAN
AGE	32	25
WINS	44	40
KNOCKOUTS	31	37
LOSSES	2	0
HEIGHT	6'3"	6'3"
WEIGHT	216.5 lbs. (98.2 kg)	220 lbs. (99.8 kg)
REACH	80"	78.5"

"I AM THE PRETTIEST OF ALL TIME." ALI PREPARES FOR THE FIGHT.

THE BUILD-UP

ALI is the first to emerge from the dressing room, preceded by an American flag and wearing a white dressing gown with stripy African trim. The huge moustachioed man towards the front of his entourage is his younger brother RAHMAN. The other key figures are his trainer ANGELO DUNDEE, his doctor/cornerman FERDIE PACHECO, and his best friend and cheerleader DREW "BUNDINI" BROWN, who sports a natty silk jacket with Ali's name emblazoned on the back. FOREMAN will keep Ali waiting so long—about seven minutes—that his eventual appearance in a crimson-red gown is greeted with sporadic boos. Ali won't appear unsettled by the delay. He takes the opportunity to conduct the crowd's chanting of the phrase "Ali boma ye!" ("Ali, kill him!"). Foreman will be attended by his manager DICK SADLER and his trainer ARCHIE MOORE.

A rather jerky, tuneless rendition of the STAR-SPANGLED BANNER with a touch of New Orleans jazz funeral about it begins the ceremonies. Then the crowd will sing along to the ZAIREAN NATIONAL ANTHEM. A lengthy pause follows, during which gloves are put on the fighters in the ring (their hands were strapped under supervision back in their dressing rooms). Finally, referee ZACK CLAYTON will call the boxers to the centre of the ring. In the pre-fight huddle, Ali counters Foreman's death-stare with constant chatter. You won't be able to hear the words but his flashing white gumshield leaves no doubt that they are tumbling out.

As the fighters go to their corners, Ali dances, breaking off briefly to pray to Allah. Foreman will perform some highly intimidating shoulder stretches, grasping the top rope on either side of his padded corner post and sending the whole structure quivering. The message is clear: "I own this ring."

ROUND 1

Ali will come out on the offensive, dancing like he did in the mid-1960s. Significantly, he will land the first punch. Indeed, he will land most of the punches—stinging, cat-like jabs invariably aimed at his opponent's head.

It will quickly become apparent that Ali is adopting the first of two high-risk strategies he will employ during the fight. This is LEADING WITH HIS RIGHT HAND, despite his conventional, left-foot-forward stance. His punches thus have farther to travel, giving Foreman fractionally more time to respond. Other things being equal, this would seem a suicidal tactic. But other things are not equal. Ali has the hand speed of a karate master and George is hard-wired to expect him to lead with his left.

One minute in you will see Ali pull Foreman's head downward and whisper in his ear. He will do this throughout the fight. But Foreman will advance relentlessly, attempting to corner Ali and cut him off from the relative safety of the open ring. Every time he manages this, he will unleash fearsome hooks to the body. You will be unable to help but fear for Ali's safety. But even while Foreman is pummelling his torso, Ali will be landing sharp blows on the Texan's head.

ROUND 2

The pace will drop somewhat after the frenetic opening round. Ali will now reach for the second counterintuitive tool in his locker: retreating to the ropes and inviting Foreman to hit his body at will. The unspoken deal is that he is not allowed to hit his head. Ali will enforce this with intense watchfulness, holding his gloves in front of his face in an almost airtight guard, and using the elasticity of the ropes to lean out of the way of any head punches that do get through. Ali's tactic of LYING ON THE ROPES and encouraging Foreman to hit him goads the champion into huge, curving body punches, leaving a channel in the middle, which Ali will use to pick him off with head shots almost at will. Ten seconds before the end of the round, Ali will shake his head, pantomime style, to indicate that Foreman's punches are not hurting him.

Do look out for the vigorous disagreement conducted at ringside between JOE FRAZIER, who reckons Ali will get slaughtered if he retreats to the ropes, and the American football star JIM BROWN, who will argue that he is the one landing the best shots despite being on the ropes.

ROUND 3

One minute in, Ali will show incredible hand speed, landing a sequence of COMBINATION PUNCHES. Then Foreman will launch a scary flurry of BODY SHOTS. Two minutes in, Foreman will land his best punch yet on the left side of Ali's jaw. But by the end of the round, Foreman will be feeling in front of himself with his fists with the tentativeness of a person stumbling in the dark. After the bell rings, watch Ali glare at Foreman as the two fighters separate. "I will not be cowed by you," his wide-eyed stare will say.

ROUND 4

The genius of Ali's "ROPE-A-DOPE" tactics will become ever more apparent. You will see him make no effort to avoid what has been the focus of Foreman's training—cutting him off from the ring. Instead he will yield to his opponent's intention. This will allow him to focus all his energy on throwing his own punches and absorbing Foreman's, rather than evasion. This is pure judo. When a shot does get through, much of its energy is transmitted to the ropes. Ali is tearing up the rule book, making what should be his enemy into his best friend. Throughout the round, listen out for Ali taunting Foreman: "Is that all you got?"

ROUND 5

This will be an EPIC ROUND, the best of the fight. From thirty to 120 seconds in, Ali will stay in more or less the same position on the ropes while Foreman gives him everything he's got. But, in Norman Mailer's memorable metaphor, Ali will lurch out of the way and back in with counterpunches "with all the calm of a man swinging in the rigging." Foreman won't be hurting him, just exhausting himself. Between the fifth and sixth rounds, attempts will be made to tighten the ropes, to prevent Ali from leaning back. You will notice his trainer, Angelo Dundee, screaming at the ring technicians to desist.

ROUND 6

A SLOWER ROUND, in which Ali will appear to be assessing Foreman's remaining strength, like a chef testing a sauce. He will deliver a series of unanswered left jabs towards the end of the second minute.

ROUND 7

This will be the QUIETEST ROUND of the fight. There is something zombie-like about Foreman's constant, unthinking forward motion. He looks as if he is sleepwalking. Ali is just biding his time.

ROUND 8

Ali will start the round SCORING AT WILL. Then, after twenty-five seconds, Foreman will throw a huge, missing haymaker—the momentum of which almost carries him out of the

ALI THROWS THE BIG ONE. CUE GLOBAL PANDEMONIUM.

ring. Halfway through the round, Ali will make his way to the opposite corner of the ring and stay there, dodging or absorbing Foreman's now-perfunctory punches. With fifteen seconds to go, Foreman will throw, miss, and lean over the ropes. Ali's eyes will suddenly light up. He will launch a devastating MULTIPUNCH COMBINATION, culminating in a straight right to the face, which will send Foreman wheeling to the canvas like a broken helicopter. Absolute pandemonium breaks out in the ring. DAVID FROST, commentating at the ringside, will squeak: "This is the most joyous scene ever in the history of boxing."

THE AFTERMATH—AND DEPARTURE

After his victory, Ali will collapse briefly, but you probably won't see this, as he will be surrounded by leaping and rejoicing members of his corner, as well as fans and hangers on. You may also notice DON KING floating towards Ali seemingly without moving his feet. Eventually several white-helmeted Zairean policemen will pile into the ring as well, not least to avoid missing out on the moment.

Kinshasa will be partying and celebrating long into the night and the next day. Feel free to join in, but do bear in mind that your DEPARTURE is scheduled for midnight today back at La Gare Central.

PART FOUR

Epic Journeys & Voyages

In Xanadu with Marco Polo

JULY 1275–FEBRUARY 1276 ✵ CHINA

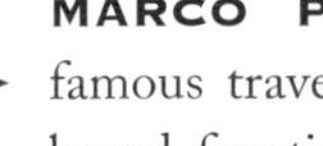

MARCO POLO IS PROBABLY THE MOST famous traveler who ever lived. His name is a global brand, fronting everything from cruise ships to clothing. But to what exactly does he owe his fame? His journey was not that arduous or hair-raising, by true explorer standards, nor was he even the first European to make the trip to the Far East. In fact, what earned him his place in posterity was the seventeen years he spent in Mongol-ruled China and at the court of Kublai Khan—the legendary Xanadu.

This trip concentrates on Polo's first six months in the great Khan's orbit, during which time you will initially stay in Xanadu (*Shangu*), the wondrous city built by Kublai as his summer residence. Polo's descriptions of this oasis of tranquillity echoed down the ages and was the inspiration for Coleridge's celebrated opium-induced dreamscape. Today, nothing at all remains of Xanadu, but you will see the city at its shimmering best.

From Xanadu, you will accompany Polo and Kublai's court on its imperial progression south to the Khan's new capital, Beijing (*Dadu*), which he has completely rebuilt, making good on the destruction wrought by his grandfather Genghis Khan. Having settled on a name for the city after consulting the *I-Ching*, he has redesigned it and constructed a stunning array of palaces and parks.

Here, at the centre of the largest land empire in history—encompassing Russia, Persia, Central Asia, and China—you will be immersed in the polyglot mass of humanity drawn to Beijing. You will witness the elaborate rituals of the court, revel in a wide variety of entertainments, and participate in the two most extravagant ceremonies in the Imperial calendar: KUBLAI'S BIRTHDAY and MONGOLIAN NEW YEAR, a massive and magnificent banquet of great symbolic and cultural significance.

BRIEFING: KHAN AND THE POLOS

Having outmanoeuvred his rivals, Kublai was proclaimed the Great Khan of the Mongol Empire in 1260. Though his realm technically included the western swathes of Mongol territory, his focus would be on China and its immediate neighbors. He added Tibet, Korea, and Yunnan to his portfolio, then turned his attention to the unfinished business of conquering the whole of China; Genghis had brought the northern half, ruled by the Jin dynasty, under his control, but the heavily populated and prosperous southern regions, home of the Song dynasty, remained independent. A long campaign, involving hundreds of ships, thousands of men, and some of the most sophisticated siege weaponry ever built, eventually ended the Song's resistance.

Word of Kublai's deeds filtered its way along the Silk Road trade routes back to Europe. Keen to discover more, Marco Polo's father, Niccolo, and uncle, Maffeo, who had been wandering the Caucasus in search of business opportunities, made their way slowly to China.

As luck would have it, Kublai was looking to develop a Christian presence in China—his mother was a Nestorian—alongside the Taoists, Buddhists, and Muslims that competed for his attention. Kublai understood that if he favoured one faith above the others he was inviting trouble: the followers of those given secondary status might plot against him, while the devotees of the primary religion might prove strong enough to act as an alternative source of power. Consequently, keen to add another church into the mix, he welcomed the Venetian travelers with open arms.

Having enjoyed Kublai's hospitality, Polo's father and uncle were dispatched homewards with a mission: to return as soon as possible with a hundred high-ranking emissaries of the Pope and to bring valued Christian relics. When they finally got back to Venice after their extended round trip, Polo was only fifteen years old, but ready and eager to join them. Two years later, in September 1271, the group set off with a letter from the Pope and some holy oil from the sepulchre of Christ, purloined by the Venetians during a stopover in Jerusalem.

The journey to China took the party around three years. Their start had been delayed by papal politics, and their journey involved lengthy diversions across brutal deserts and snow-covered mountains to avoid hostile locals. But in 1275 they reached the borders of Kublai's territory, and from then on it was plain sailing. Kublai had the party accompanied by a large Mongol contingent and furnished them with a gold *piazi*—a travel pass that ensured they completed the journey unmolested, arriving at the emperor's court in the summer of 1275.

MARCO POLO IN TARTAR COSTUME.

THE TRIP

You will meet up with the Polos at one of the many relay stations (*yamb*) that appear at regular intervals across the Khan's territory and act as both crucial nodes in his communications infrastructure and refuelling stations. Your *yamb* is at XUANHA, a few days' travel from Xanadu, located on a level basin between mountains, and with well-appointed, comfortable lodgings. Here you will take on supplies and around 300 fresh horses, mostly for the military escort that will now accompany you on this first leg of your journey.

It is at Xuanha that you will become part of the Polo entourage, and you will go on to play a range of supporting roles over the course of the trip. Expect some light menial duties and a certain amount of bowing and scraping. Nevertheless, you will share the exclusive access of the Polos to Kublai's court; this privilege will surely outweigh any unpleasantness you may encounter as a domestic helper.

At the same time, both in Xanadu and Beijing, you will be able to merge seamlessly into the small army of staff and attendants needed to keep the wheels of Empire turning—cooks, artisans working in gold, silver, porcelain, and textiles, servants, managers, entertainment specialists, historians, translators, interpreters, astronomers, doctors, librarians, shrine-keepers, musicians, architects—and lose yourself amongst them, giving you the freedom to wander and explore as you choose.

During the summer months you will wear pyjama-like CLOTHING; for the winter, you will be supplied with coats made of sheepskin, furs, and animal pelts, thick boots, and the ubiquitous Mongol furry hats. You will also be provided with enough cash to cover your expenses for the duration of the trip. Kublai is very keen on PAPER MONEY, which has been in circulation in China for several hundred years, and has introduced three new notes—two backed by silver, one by silk. These currencies are made from paper that comes from the beaten inner bark of mulberry trees, and is printed using carved wooden blocks. With a wad of these notes, you will have plenty in the way of spending money.

XANADU (SHANGU)

You will approach the city of XANADU through open grasslands, passing a series of hills marked with shrines of sacred stones (*obo*). Notice the steady build-up of traffic—mostly carts pulled by bullocks or yaks—on the many tracks converging towards it. Up to 500 vehicles a day travel this route in order to supply Xanadu's 120,000 citizens.

You will be sticking to the main thoroughfare, the ROYAL ROAD, from which you will marvel at the thousands of pure white horses, the most prized animals that Kublai possesses, ranging over the plain, and a similar number of musk deer, much coveted for their distinctive perfume. Your route then takes you into the OUTER CITY, with its dense conglomeration of simple mud-brick houses, traders, and thousands of cooking stalls wafting tantalising aromas.

Passing on, you reach Xanadu proper. The site for the city was chosen by Kublai's foremost town planner, Liu Bingzhong, and members of his Golden Lotus Advisory Group. As the area was previously known as the Dragon's Ridge (*Lung Gang*), spells were cast to evict said dragon and a magical iron pennant was raised to prevent its return.

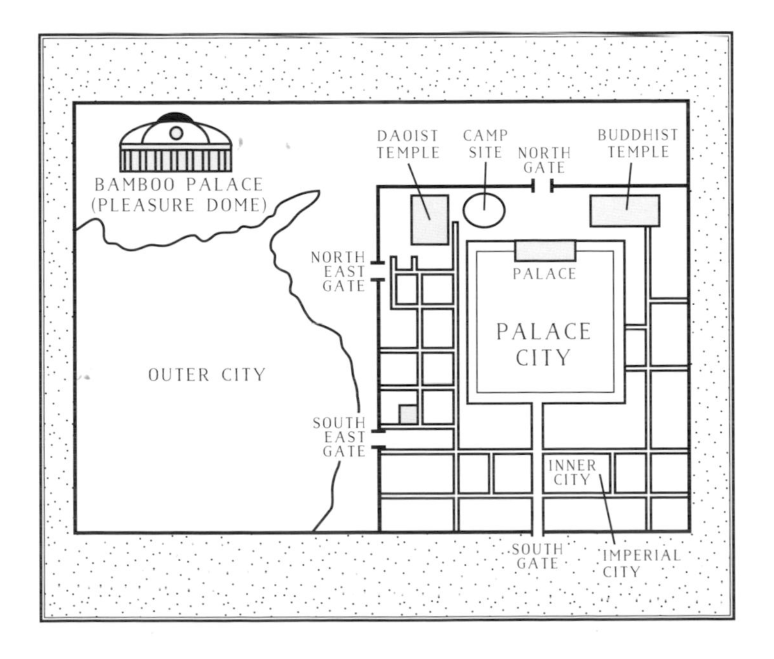

Xanadu divides into three sections, all squares, with a twenty-five-foot-high outer wall, marked by bastions, towers, and large gates, that stretches for more than a mile on each side. Above the walls, you will glimpse a mass of roofs with blue, green, and red tiles glinting in the sun. Once inside, your senses will be assaulted by another mass of street traders peddling their wares amongst densely packed dwellings. Be careful where you tread. The very poorest live in pits dug in the ground with planks and dry grass for roofs. Step too heavy and you might find yourself in somebody's living room!

Making your way across the city, check out the WEST GATE MARKET, which does a brisk trade in horses, sheep, cows, and even slaves. Another attraction worth visiting is the NORTH YARD, a kind of zoo featuring pumas, lions, eagles, and other rare animals.

Your focus, however, will be on the inner-square walled section of Xanadu that contains the IMPERIAL CITY. You will reach it via the South Gate, or GATE OF SPLENDOUR. After crossing a moat, you will walk along a road lined with houses for about half a mile to a second wall and a second gate that opens into an area of the Imperial City, laid out in grids, that includes sub-palaces for Kublai's family and staff, temples, government buildings, and the houses of officials.

Finally, once you are over another moat, you will be inside the last and most important square: the PALACE CITY, with its series of splendid tented pavilions, their names testament to their imposing stature and beguiling aura—CRYSTAL, AUSPICIOUSNESS, WISDOM, CHARITY, FRAGRANCE, and CONTROLLING HEAVEN. Then take a deep breath, as rising before you is the ROYAL PALACE, a two-storey building on a 350-foot-high brick-faced platform, much of it made from white marble especially imported by Kublai.

As befits the ruler of most of the known world, Kublai tends to keep his visitors waiting, and you and the Polos will have to linger a while for your first audience and your first glimpse inside his Imperial Palace. Having changed into the special footwear, beautiful slippers of white leather that all visitors are required to wear, you'll be escorted into a labyrinth of more than 120 rooms—two wings, a central courtyard, and its grand Pavilion of Great Peace—which the poet Zhou Boqi describes as "rafter upon rafter, the storied pavilion reaches the azure sky, a picture painted in gold floating atop seven precious pillars."

You will then follow the Polos into Kublai's chamber and lay your eyes on the Great Khan—who is short, stocky, and shows signs of the gout that will plague his later years. He will be dressed in glorious robes of gold laced with intricate patterns, his black-eyed gaze scrutinising your every move.

You and the Polos should prostrate yourselves before his majesty, worried that the failure to bring along a bevy of Papal officials will anger him. Fear not. Kublai is delighted to see the Polo family, and welcomes them and, by association, you, into the bosom of his court.

LIFE IN XANADU

You will be staying in a GER (or yurt), the traditional Mongolian habitat, on a camp site for VIPs and merchants near the East Gate. Your *ger* has a wooden circular frame, covered in felt made from sheep's wool, with a frame of expanding lattice-wall sections. The interior will also be lined with felt and decorated with patterns that represent savage beasts, the five elements (fire, earth, water, metal, and wood), plus geometric patterns such as the continuous hammer (*alkhan kee*).

Considering the nomadic lifestyle of the Mongols, it'll come as no surprise that their toilet arrangements are pretty rudimentary: expect to find yourself squatting above a stinking hole in the ground. But, on the bright side, toilet paper is available—the Chinese have been using it for 600 years. Note that these open latrines are thought to harbour ghosts. To combat the presence of malicious spirits, the Ruler of Heaven (China's main deity) blessed these latrines with their very own god, the Purple Maiden. Legend has it that a first wife, jealous of her husband's second wife, murdered her rival by shoving her into a communal shit pit. The Ruler of Heaven then took pity on her poor befouled soul and made her divine.

Toilet issues aside, the time spent in your *ger* will allow you to appreciate the alfresco simplicity and back-to-nature vitality of the

Mongol way of life. With warm weather and long hours of daylight, it will be a refreshing pleasure to stay under canvas, dining out every night under the panoramic, star-speckled sky.

EATING AND DRINKING

Though most of the city's population is Chinese, Kublai is determined that Mongolian culture will predominate at Xanadu. The life-blood of their diet is the mildly alcoholic drink *airag*, derived from MARE'S MILK which has been poured into a huge sack and churned with a club before turning sour and fermenting, and then churned continually until butter can be extracted from it. You better get a taste for it quickly, as it will be offered to you constantly. The mare's milk is also boiled to get clotted cream (*orom*), with the remains processed into cheese (*byaslag*). This emphasis on dairy produce may present a challenge for those with a lactose intolerance.

As for other foods, the Mongol's most popular delicacies include steamed or boiled minced-meat DUMPLINGS (*buuz*) and MUTTON cooked over hot stones. Broths are also common fare, such as BORBI SOUP, concocted by reducing thirty or so sheep bones in a bucket of water.

There will be more sophisticated dishes on offer that combine ingredients and culinary techniques from other parts of Kublai's realm, such as RUSSIAN OLIVE SOUP, consisting of one leg of mutton, five cardamom seeds, and shelled chick peas, which are boiled and strained before Russian olives, sliced sheep thorax, Chinese cabbage, or nettle leaf are added to the mix.

Another option is butter-skin YUGBA, which uses a fine cut of mutton, sheep's fat, sheep's tail, mandarin orange peel, sprouting ginger, salt, sauce, and spices, while the skins are from a blend of vegetable oil, rice flour, and wheat flour. You will also come across NOODLE DISHES with mutton, egg, ginger, sheep intestines, and mushrooms in a broth seasoned with pepper, salt, and vinegar.

VEGETARIANS may find Mongol fare a struggle. Your best bet is to stick to noodle dishes, with the meat extracted, or rice dishes that use curds or raisins for flavour. Vegans may prefer to stay at home.

HUNTING

One of the highlights of your time in Xanadu will be the opportunity to enjoy Kublai's private GRASSLAND PARK, just to the northwest of the city. While its fountains, brooks, immaculate lawns, and hidden groves summon up visions of paradise, the real purpose of the park is HUNTING, the defining activity of

KUBLAI KHAN HUNTING WITH A GYRO-FALCON—A MINIATURE FROM THE FIFTEENTH-CENTURY *LIVRE DES MERVEILLES DU MONDE*.

Mongolian culture. Kublai's park is well stocked with deer, hares, rabbits, and birds. Much of the hunting is done on horseback, using bows and arrows; this is a tricky art and will be impossible to master without some equestrian experience.

Kublai also employs a range of hunting dogs—mastiffs, greyhounds, and retrievers. But most spectacular of all is the massive GYRO-FALCON. Bred mainly in Korea and Manchuria, this sub-Arctic bird is either white with black flecks or pure white, and has a massive wingspan. You will watch open-mouthed as this immensely powerful falcon brings down a deer by battering it round the head and eyes with its wings. Equally awe-inspiring, indeed a bit terrifying, will be the sight of Kublai letting his PET

CHEETAHS loose. Marvel at the speed and agility of one of nature's most sublime predators as it pursues its prey.

Another not-to-be-missed park attraction is Kublai's BAMBOO PALACE, also known as the PLEASURE DOME. This regal *ger*, where the Khan conducts much of his official business in lavish luxury—with animal skins for a carpet—is on a scale unimaginable to his forefathers and required a considerable feat of engineering to make it large enough for his purposes while still being portable. The *ger* is secured in place by 200 silken cords, and rests on top of pillars, each with a great golden dragon. The whole thing, despite its size, seems to float in the air, defying the laws of physics, rendering it somewhat fantastical to the eye, like something wrought out of legend.

WORSHIP

Xanadu is host to a multitude of BUDDHIST AND TAOIST TEMPLES, the biggest of them attended by several hundred monks. They are in the pagoda style that is still with us today, with multiple roof sections, bright colors, and dragon motifs, while inside the ornate interiors are golden statues of the gods and incense burning continually. Feel free to enter and take part in ceremonies that have remained practically unaltered for thousands of years.

Though Kublai is careful not to over-privilege either religion, he has chosen to adopt MAHAKALA, the Buddhist god of war, as his patron saint; its black face, fierce, blazing eyes, fixed snarl, wild yellow hair, and headdress of skulls are an appropriate image for a serial conqueror. However, this does not mean Kublai has deserted his own spiritual roots. At Xanadu, MONGOLIAN SPIRITUAL PRACTICES will be the most frequently seen.

Based on a circular concept of time, this is a pantheistic faith, particular animals acting as totems or symbolic ancestors, the most famous being the BLUE WOLF and RED DEER, the mythical progenitors of the Mongols. Ruling over their universe are the SKY GODS (*tengri*), the foremost being ETERNAL BLUE HEAVEN, with the sun (fire) and moon (water) for eyes. Beneath him are 99 other male divinities, 55 of them benevolent (white), and 44 terrifying (black). As an equal-opportunity religion, there are 77 female divinities (*natigai*), governed by the ALMIGHTY EARTH MOTHER.

This celestial pantheon will be available to you through the ritual machinations of the SHAMANS. Gifted with psychic powers (*hii*) and the ability to travel between the worlds of the living and dead, human and divine, man and animal, they can also access three categories of ancestral spirits: Lord Spirits, Protector Spirits, and Guardian Spirits. During their rituals, you will be transported by mysterious chants and the incessant rhythmic beating of a one-sided hand-held drum (*tuur*). You will notice that the shaman wears an amulet (*dalbuur*) and a metallic circular mirror (*toil*), which not only acts as armour to deflect spirit attacks but also absorbs energy from the universe. To intensify the experience you will be offered alcohol and tobacco, and be expected to inhale the fumes from burning juniper, producing a mildly hallucinogenic high. Suspend your disbelief, open your mind, go with the flow, and you will find yourself traveling through time and space.

DEPARTURE AND IMPERIAL PROGRESS

Your stay in Xanadu will finish on AUGUST 28th, the traditional end of summer, when Kublai will depart south for Beijing. This auspicious day is marked by a LEAVING CEREMONY. You will witness Kublai sprinkling milk taken from his white mares onto the ground by hand to honour the gods, followed by the mystic men, who will drink the sacred milk that is poured out especially for them before they offer their blessings to Eternal Blue Heaven.

This ceremony complete, you will begin the journey to Beijing along with a vast train of carriages and carts filled with hundreds of Kublai's staff. Covering 20 km a day, it'll take you roughly three weeks to reach your destination. En route, you will be stopping at a string of towns whose only function is to welcome Kublai once a year. The first port of call is HANGZHOU, followed by LIANG TAI and CHAGAN NOR—known as the White Lake, and famous for its swans, partridges, pheasants, and cranes. Then onto ZHONG DU, ZHANGBEI, ZHANGJIAKU, XUANHU, TUMU, and KHARA-BALGASAN (Black City), finally dropping down through a gorge between mountains to the plains.

From here there are thirty miles of open ground between the mountains and the city of DADU (modern Beijing). You will approach it on a road that has been specially cleared and smoothed over by labourers.

DADU (KHANBALIQ)

The Imperial Progress will take you through the suburbs of DADU, or KHANBALIQ as it has been renamed by Kublai. This is the city that will become Beijing, and already it is a massive conurbation with a sea of houses and little plots of land for vegetables and other crops, alongside larger houses and inns reserved for foreigners and merchants. You will immediately be struck by the thick haze of smoke. This is the product of constantly burning FUNERAL PYRES; Kublai has outlawed the burial of bodies within the main city. You will also be struck by the multitudes of prostitutes hustling for business; Kublai banned them as well from his new capital.

THE SUBURBS

The suburbs form a thrillingly cosmopolitan and multicultural township, where you will rub shoulders with Chinese, Mongols, Turks, Arabs, and Indians, haggle with merchants offering silks, spices, jewels, and pearls, be entertained by buskers, and tempted by the STREET FOOD being prepared on every corner. This is largely of Chinese origin and based on the northern school of cooking, Lu, and will not be that different from a takeaway back home, with similar ingredients, spices, and seasoning. And, of course, there is more tea than you could ever possibly drink.

This is also the area to visit for a night on the town. The numerous INNS offer rice wines and local beers and the opportunity to participate in games such as MAHJONG and indulge in a spot of GAMBLING. Playing cards have been around in China since the early ninth century, and you may have an edge at the tables, as one of the most popular card games is very similar to poker. Otherwise, you might want to try Chinese dominoes, using tiles (*kwat pai*) with symbols/characters on them and many-sided dice, or Chinese chess, with pieces that resemble elephants. Backgammon is available too, but may prove an expensive option, as it tends to attract high-spending high-rollers with cash to burn.

THE CITY

Beijing is organised on the same principles as Xanadu—once again the work of Liu Bingzhong, assisted by an architect from Central Asia, Ikhtiyar al-Din. It takes the form of a palace within an inner city, within an outer city, sporting huge walls all around, then a second wall concealing the Imperial City, then a third one shielding the Palace.

When your procession reaches the thirty-foot-high whitewashed main wall, it will turn south and proceed four miles to the South Gate, or EMPEROR'S GATE, which will usher you into the Imperial City, with streets laid out in grids, raised pavements, good drainage, and whole blocks owned by the wealthiest and most prominent families.

DADU LODGINGS

In the Imperial City, you will notice a tented zone similar to the one in Xanadu, with grand *gers* for VIPs and smaller ones for officials and craftsmen, and for weapons stores. You may pitch up here if you wish; however, given that conditions will turn distinctly chilly and wet over the coming months, you may prefer to abandon the world of the *ger* and opt to take up residence with your masters in the well-appointed and spacious mansions they inhabit while in Beijing. Not for the Polos the lodging houses of the suburbs. Kublai had already taken a shine to them and considers their advice invaluable; he wants them close by at the centre of his realm. Aside from being more comfortable, you will also have access to better toilet facilities: a sheltered wooden structure with a stone seat to perch on and an armrest, set over a pit in the ground that has running water to flush it out, plus highest-quality toilet paper, soft and perfumed.

INNER SANCTUM

Accompanying the Polos, you will make frequent visits to the PALACE itself. You will cross a moat with a three-arched marble bridge, stroll along three passageways to three gates with towers and five doorways, each one gaining you entrance to the IMPERIAL PALACE and Kublai's private gardens, with lakes full of fish.

The centrepiece of Kublai's conception is the resurrected JADE ISLAND, which had been left to rot by Genghis Khan. Kublai has rebuilt the bridge that leads to it, landscaped the slopes with rare trees, added winding staircases, temples, and pavilions with evocative names—Golden Dew, Jade Rainbow, Inviting Happiness, Everlasting Harmony—and a main temple that sits atop the island, the grand WHITE PAGODA. Here you can spend many blissful hours in quiet contemplation of the island's outstanding beauty.

Then there is the spectacular NEW PALACE, a single-storey structure raised above the ground by marble terraces and staircases, lined with varnished woodwork and sculptures of dragons and other potent animals, and famous warriors. Inside are seven halls, plus dozens of rooms, chambers, treasure troves, offices, and apartments for Kublai's four wives and many concubines. Everywhere you go, you will see fearsome-looking heavily armed men. These are the elite warriors that comprise Kublai's personal bodyguard—it's best to stay on their good side.

Within the sprawling building there is a MINI-PALACE where Kublai holds court from a huge bed inlaid with gold and jade. Next to the bed is a huge jade urn weighing around four tons, filled with 600 gallons of wine. Perhaps at no point in your stay will you be as impressed with Kublai's majesty and the authority that emanates from his person as he reclines on his bed, completely at ease with the power that he wields so naturally, totally secure at the heart of his vast empire as he watches his servants dispense wine to his grateful guests.

A NIGHT AT THE THEATER

In accordance with Kublai's wishes, Chinese culture is to the fore in Beijing, and this means, above all else, THEATER. As a result, you will get to see everything from large-scale productions to smaller shows performed on movable stages. These events will combine music, singing, poetry, pantomime, dance, and acrobatics, punctuated by comedic interludes, running gags, and audience interaction.

Many of the more formal four-act plays will feature love stories with strong female characters in the leading role (*tan*), often reworking well-known plots and myths. Originality is not at a premium; much of the fun for the audience comes from seeing their favourite bits from other shows, and stock characters such as the clown (*ch'ou*), the comic villain (*chou*), and the malevolent female (*ch'a tan*). Another aspect of these performances that will be familiar to the spectators is the mimes: choreographed and stylised movements from a set repertoire of gestures and expressions: for example, *entering lovesick*, *having a coughing fit*, and *rubbing eyes in disbelief*.

The key element that binds the plays together is music. At first you may struggle to appreciate the atonal and apparently discordant sounds on offer, but be patient, as you will come to enjoy and admire the elegant, complex harmonies and the skill and dexterity of the musicians playing a wide range of string, woodwind, and percussion instruments to produce the eight tones (*bayin*). There are various types of zither, harp, and lute, which can be plucked, bowed, or struck. The woodwind section will consist of bamboo flutes and pipes, as well as conch shells (*hailou*), while there is a whole panoply of percussion—stone chimes, metal bells, cymbals (*bo*), gongs (*lou*), and an array of skin drums. Expertly deployed, they produce sounds of great subtlety and passionate intensity.

BANQUETING

During your stay you will be lucky enough to take part in two of the greatest feasts of the year. First up is the celebration of KUBLAI'S BIRTHDAY on September 23rd, which takes place in the 6,000-capacity HALL OF GREAT BRILLIANCE. You will join the other commoners who will be sitting on the floor gawping at Kublai's robes of beaten gold and thousands of nobles and attendants dressed in similar garb.

Take special care when entering the hall to STEP OVER THE THRESHOLD without your feet touching it. This custom relates to life in the *ger*: to touch its threshold with your feet by accident is considered a bad omen; to do it deliberately, a terrible insult. If you happen to trip up as you go into the hall, the massive guards by the doorway will beat you with rods and strip you of your outer garments; oddly, the only mitigating circumstance is if you are too drunk to walk straight.

The other magnificent banquet you will attend is the WHITE FEAST in early February. This celebrates Mongolian New Year and marks the end of your trip. The festivities begin solemnly as you and the assembled guests touch your foreheads to the floor four times, followed by a brief song and a prayer. Then Kublai takes his place at the high table with his head wife, Chabi, known for her wisdom and temperance, next to him, while the princes and their wives sit on a lower platform level with his feet. At the centre of proceedings is a huge decorated buffet table and a vast golden wine bowl from which butlers draw wine into golden jugs. The very first cup is for Kublai; once he takes a sip, everyone else can. The same ritual is repeated with the food, which will be mostly the meat pastries (*buzz*) you will remember from Xanadu. You will notice that those serving Kublai have their mouths and

AT THE BANQUET YOU'LL BE AWARE OF EMPRESS CHABI, KUBLAI'S FAVOURITE WIFE AND SPECIAL ADVISOR. MONGOLIAN-BORN, SHE HAS BEEN INSTRUMENTAL IN THE RETURN OF THE POLOS TO KUBLAI'S COURT.

noses wrapped in veils of silk and gold, so neither their breath nor smell contaminates his dinner.

Throughout, an orchestra will perform MONGOLIAN MUSIC. Similar to the Chinese tradition, the musicians will play either string, woodwind, or percussion. The most common instruments are the horse-head fiddle (*morin khuuur*); a three-stringed strummed lute with a long neck that sounds like a banjo (*shanz*); flutes (*tsuur*); oboes (*everburee*); and a richly ornamented metal trumpet with a brass mouthpiece (*bishguur*). Percussion will include small bells (*damar*) and frame drums (*tuur*). The most evocative and startling element is the THROAT SINGING (*hoomii*) that produces two distinct sounds at the same time: a low guttural drone alongside higher melodic notes. Tune into the ecstatic vibrations of the throat singer's trance-inducing sounds, and it will stir your soul and animate your spirit.

Once the eating is done, there will be a FLOOR SHOW featuring cabaret, acrobats, jugglers, and conjurors. Pace yourself. The carousing and merrymaking will go on well into the following day.

DEPARTURE

While your masters are still in bed sleeping or nursing their hangovers, slip quietly away and head for Jade Island. Once there, you will DEPART from the TEMPLE OF EVERLASTING HARMONY.

Captain Cook's First Epic Voyage

AUGUST 26, 1768–JULY 12, 1771

CAPTAIN JAMES COOK MADE THREE EPIC voyages to the Pacific—and we are delighted to offer the first and best-known, a three-year odyssey that included the first accurate charting of the New Zealand coastline and the "discovery" of Australia. As a member of Cook's crew you will experience tremendous highs and devastating lows, and your strength of character and physical stamina will be tested to the limits. But the rewards lie in stunning, unspoiled landscapes and untamed shores; you will marvel at wildlife of all kinds, pit yourself against the elements and the raw power of the sea, and stand on deck to watch the sun set over the Pacific.

Setting off from PLYMOUTH, England, your journey will see you cross the Atlantic, with stopoffs in Madeira and Rio de Janeiro; negotiate the perilous passage round Cape Horn into the Pacific; spend several months in Tahiti, the island paradise with its legendarily welcoming inhabitants; then sail on to New Zealand, with its very much less welcoming locals, where you will circumnavigate both the North and South Islands before heading towards Australia. There you will have a close shave with the Great Barrier Reef and make the first European contact with the Aborigines. Having claimed this new continent for the British Empire, you will make a gruelling journey home.

BRIEFING: A SCIENTIFIC EXPEDITION

Cook was not the first European to venture into the Pacific, and at the time of his first voyage there is a generally held view among Europe's geographers that a new continent—*Terra Australis Incognita*—is waiting to be found. This, of course, is of interest not just to Britain, the world's dominant sea power, but to other European contenders for global empire—the French, Dutch, Spanish, and Portuguese.

A scientific expedition has been judged to be the best way of furthering British imperial ambitions without raising the suspicions of its rivals. Thus, in February 1768 the ROYAL SOCIETY petitioned George III for cash to fund a voyage to the Pacific to observe the TRANSIT OF VENUS, the rare stellar event that, if measured correctly, can reveal the distance between Venus (and other planets) and the sun.

CHARLES GREEN, assistant to the Astronomer Royal, has been chosen for the star-mapping, and to boost the expedition's scientific credentials a team of botanists has been enlisted. They are led by JOSEPH BANKS, a very wealthy man who has devoted himself to the system of classifying flora and fauna developed by the Swede Carl Linnaeus. Banks has engaged his pupil DANIEL SOLANDER and a Finnish naturalist, HERMAN SPÖRING, as assistants. By the end of the voyage, his team will have collected 30,000 botanical specimens, including 110 new *genera* and 1,300 new species. Two artists, SYDNEY PARKINSON and ALEXANDER BUCHAN, have been invited along to record the trip.

The forty-year-old JAMES COOK is from a humble Yorkshire farming background and has been appointed captain for his skill as a cartographer. He had helped survey and chart the Gulf of St Lawrence in Newfoundland during the Seven Years' War (1756–63). The ship chosen for him by the royal Society, the ENDEAVOUR, is a one-time coal ship—a sturdy vessel weighing 368 tons. It is 97 feet long and 29 feet wide, with a broad, flat bow, square stern, and deep hold. In preparation for the trip, its hull has been sheathed and caulked, and a third internal deck added for extra cabins and storage space. It also houses a longboat and several smaller boats, while ten four-pounder cannons and twelve swivel guns give it some teeth.

The crew comprises fourteen officers, twelve Royal Marines, eight servants (two of them African and attached to Banks), and around sixty sailors from all over Britain. Adding to the ship's compliment is a GREYHOUND for hunting, three CATS for killing rats, and a GOAT, which is more a mascot than anything else.

As is customary before departure, anybody suffering doubts is given the chance to quit. Eighteen men vote with their feet. Replacing them at such short notice is the job of the press gangs, part of the IMPRESS SERVICE, and this is how you will be joined up.

THE TRIP

You will ARRIVE on August 18, 1768, in TURNCHAPEL, PLYMOUTH, an area of dark, narrow, cobbled streets right by the harbourmouth. Violent, seedy, full of brothels and bars, it is fertile ground for the Impress Service. Take a pew in the BORINGDON ARMS, a dockside pub, and wait to be approached. Each press gang of twelve men is led by an Impress Officer who is paid by results. He will give you a choice between being taken by force or becoming a VOLUNTEER. The latter is, of course, much the best option. As a "volunteer," you'll get paid the king's shilling, in advance, and will receive better treatment.

You will spend the majority of the trip with your fellow crew members. It is therefore essential that you adapt quickly to their way of life. If you think you are going to hang out with Cook, or indeed any of the other "gentlemen" on board, then you will be sorely disappointed. An unbridgeable gap exists between you and them. There will be some interaction concerning the day-to-day doings of the ship, but ultimately your place is with the crew. Together you will be responsible for transporting Cook round the world; without you the ship won't get very far.

Some experience of being at sea—and sailing—would obviously be useful, as would familiarity with the layout of ships and the terms used to govern them. Such knowledge will earn you the title of ABLE-BODIED SEAMAN. However, the royal Navy doesn't mind if you are a nautical novice; you will simply be referred to as a LANDSMAN. Either way, you will not (unless you are particularly fit and well suited to such duties) be expected to perform the most challenging tasks, such as going aloft to unfurl the sails and untie the topgallants. These dangerous and vertigo-inducing jobs are normally reserved for the youngest and most agile sailors, known as the UPPER YARD'S MEN.

Instead, you will be on DECK DUTIES, hauling sheets, halliards, and braces. You will be equipped with a knife for cutting lines that you will carry sheathed on your back. Otherwise, your main duty will be going on WATCH; you will be assigned a specific part of the ship to monitor, and each shift will last eight hours. However, rather than doing three consecutive nights in a row (standard practice at sea), you will only do two, thanks to Cook's largesse.

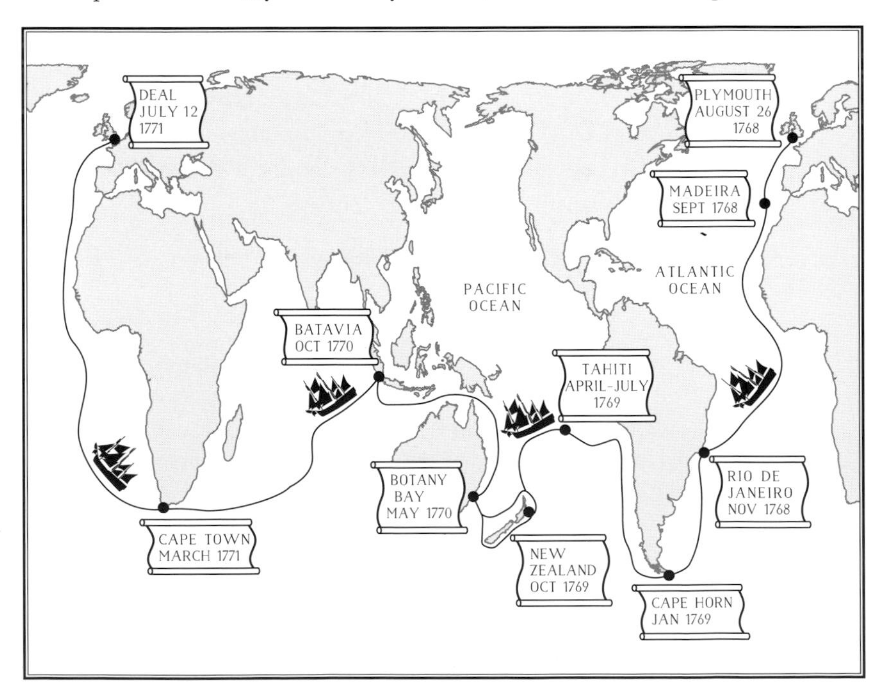

ON BOARD SHIP: PRACTICALITIES

Your BASIC KIT will consist of a buttoned single-breasted waistcoat (*weskit*), long-sleeved cotton or linen shirts with short cuffs and a single-button, baggy striped *pantaloons* of canvas to cover your knee breeches (*slops*), and heavy trousers and *fearnought jackets* for when it gets cold. Hats are a staple: either variations on the triangular-shaped tricorn, turned up so the brim diverts rainwater away from your face, or a brown woollen-knitted *Monmouth cap*. Your shoes will be black leather with a brass buckle, though much of the time bare feet will suffice. You may wish to accessorise with a bandana, often in black to mask the ingrained dirt, and a brightly colored neckerchief.

FOOD

On a voyage like this, your well-being, both mental and physical, will hinge on the quantity, quality, and variety of FOOD on offer: your morale and that of the crew will revolve round your stomachs. The contents of your diet are largely determined by Cook's desire to prevent scurvy, the deadly wasting disease so prevalent amongst sailors.

Drawing on the latest research, especially James Lind's 1753 *Treatise of the Scurvy*, Cook's selection of anti-scurvy food will include a hundred pounds of *portable soup* (from dried vegetables), sugar, sago, marmalade of carrots, ground up orchid roots (salop), vinegar, mustard, malt, raw onion, concentrated lemon juice, shore greens, and 7,860 pounds of pickled sauerkraut. *Endeavour* will leave Plymouth carrying, in addition, 6,000 pieces of pork and 4,000 of beef, nine tons of bread, five tons of flour, one ton of raisins, cheese, salt, peas, oil, and oatmeal, plus seventeen sheep, four pigs, and twenty-four assorted poultry.

You will have three-quarters of an hour for BREAKFAST and an hour and a half for LUNCH, consuming roughly 4,500 calories per day. Your weekly ration will contain four pounds of salt beef, two pounds of salt pork, three pints of cereal grain, six ounces of butter, and twelve ounces of cheese. Mondays, Wednesdays, and Fridays are meat-free; you will be served pease pudding and onions instead. SHIP'S BISCUITS, pretty vile and often infested with insects, will accompany every meal.

Over the course of the journey, fresh water will be procured at regular intervals and you will be able to gorge on produce sourced from the many islands you visit; coconuts, yams, bananas, sweet potatoes, yellow apples, sugar cane, and plantains. FISH AND SHELLFISH will also make a welcome change. Savour oysters and lobsters fresh from New Zealand's rocky inlets; feast on huge stingrays plucked from the waters off Australia; and tuck into giant turtles, popular not only for their succulent meat but also for the sport to be had diving for them.

Sea birds will also land on your plate, most notably ALBATROSS. Though the crew consider this strange, ungainly bird the sacred embodiment of the spirits of dead sailors, this doesn't stop the botanist Joseph Banks shooting one. The albatross will then be served up in a savoury sauce, having been skinned, soaked in salt water, parboiled, and stewed until tender.

DRINK

After food, ALCOHOL is the mainstay of the crew. On board are 250 barrels of beer, 44 barrels of brandy, and seventeen barrels of rum. Out of this, you will be given a DAILY RATION of a gallon of beer or a pint of grog—rum diluted with water; sometimes the crew will mix beer with spirits to make a concoction known as *flip*. Not surprisingly, drunkenness is endemic and you will need to be able to handle your liquor without losing control; much of the trouble on the ship will be caused by too much alcohol. One crew member will die after drinking three and a half pints of rum, while four will be flogged for stealing rum.

If you are a teetotaler, for whatever reason, the best way to explain your abstinence is to claim serious religious convictions—preferably Protestant.

Aside from drinking, recreational activities will include singing SEA SHANTIES and GAMBLING on pretty much anything from arm wrestling contests to cockroach racing. For those so inclined, there will be a RELIGIOUS SERVICE every Sunday morning.

ACCOMMODATION

The quarters where you sleep will be on the lower deck. Your HAMMOCKS are a mere fourteen inches wide with hardly any gap between you and your neighbors, while the ceiling is only four feet high. Anybody over six feet is going to struggle, as is anyone who suffers from even mild claustrophobia.

Going to the toilet will present its own unique challenge. The crew's LATRINE is simply a hole cut in a long plank extending out from the bow of the ship. Privacy and retaining your dignity will be the least of your worries; you will be focused on not falling off the plank. In rough seas, you may prefer just to hold it in.

You will wash yourself regularly using a large bucket full of cold water. Clothes will also be soaked in it and scrubbed with soft soap made from ash lye and animal fat. Your hammock and bedding will be rolled out and subjected to the same treatment. All these items will then be hung out on deck to dry in the sun and the wind.

DISCIPLINE

Once a week you will gather on deck for a reading of the ARTICLES OF WAR. First drawn up in 1652 and last updated in 1757, this has thirty-six sections governing matters of discipline and punishment on board ship, listing crimes either meriting the death penalty or "such punishment...as the nature and degree of the offence shall deserve," which generally means getting FLOGGED by a cat o' nine tails. Known as the "captain's daughter," this is about two and a half feet long, with nine waxed cords of rope, each with a knot on the end. A maximum of twelve blows will be delivered to your bare back, the whip cutting into your flesh and causing considerable bleeding. Afterwards, salt will be rubbed into the wounds to prevent infection.

Thankfully, many of the Articles will not apply to your voyage. However, a number are worth paying close attention to. Take care not to fall foul of ARTICLE TWO, which covers "profane oaths, cursing, excretions, drunkenness, uncleanness or other scandalous actions"; ARTICLE TWENTY-THREE, which condemns fighting, quarrelling and winding up your fellow crew members (an easy trap to fall into on such a long and stressful journey); and ARTICLE TWENTY-SEVEN, which warns you not to fall asleep while on watch or slack off from your work, both errors you may unwittingly commit. The clause most open to abuse is ARTICLE THIRTY-SIX, which covers "all other crimes...not mentioned in the act," and gives the captain carte blanche to punish any misdemeanours. Luckily for you, Cook uses the lash sparingly. He is a benevolent captain who believes in treating his crew humanely. This does not mean Cook is a soft touch. He realises the necessity of maintaining discipline and will act accordingly.

However fed up you are with your diet, don't gripe about the food. ARTICLE TWENTY-ONE forbids complaints about "the unwholesomeness of the victual," something Cook takes seriously. Two of the crew will get twelve lashes for refusing to eat their ration of fresh beef.

Even if you avoid such disciplinary measures, you will have to watch the whippings, performed on deck in front of all of you, accompanied by drum rolls, sips of water for the victim and the repetitive crack of the lash.

DEATH

An unavoidable and distressing part of the trip will be the loss of fellow sailors. The first death will occur only two weeks after leaving Plymouth: one of the crew will get dragged overboard by the buoy-rope and anchor. Six months later, either deliberately or by accident, a marine will disappear over the side, while on arrival in Tahiti the artist Alex Buchan will die of a bowel disorder. Overcoming the grief of these unfortunate deaths will help prepare you for when the toll increases dramatically later on in the voyage.

CROSSING THE ATLANTIC

AUGUST 26–NOVEMBER 13, 1768

Out of Plymouth, and through much of September 1768, the *Endeavour* will be buffeted by gales and heavy rain; in the worst of the storms, one of the small boats will be washed overboard, along with a lot of poultry. But on the 13th of the month, everyone aboard will get some welcome relief during a four-day stopover at the port of FUNCHAL on the island of MADEIRA.

While the ship is being re-caulked and painted here, you will have a chance for a spot of shore leave: Funchal is a busy port, so you can expect the usual conglomeration of bars and brothels. However, there will be little time for sightseeing because by day you will be transporting fresh supplies and loading them onto the *Endeavour*: an extra twenty pounds of onions per seaman, 270 pounds of fresh beef, a live bullock, plus 1,200 gallons of beer, 1,600 of spirits, and 3,032 of wine.

On Wednesday, October 26th, you will reach the EQUATOR—"crossing the line"—and have the chance to participate in the Navy's equivalent of bungee jumping: any sailor who has never crossed the line before will be ducked in the ocean three times. Suspended above the yardarm, you will be fastened by rope to three pieces of wood—one between your legs, one between your hands, and the other above your head—hoisted as high as possible, and plunged rapidly beneath the waves before being pulled up again. You are not obliged to take part in this ceremony, but if you don't you'll forfeit four days of alcohol rations.

By the time RIO DE JANEIRO comes into view on Sunday, November 13th, you will be chomping at the bit to sample its many delights. However, the Viceroy of Brazil, Dom Antonio Rolim de Moura, has other ideas. Unimpressed by the *Endeavour*'s scientific credentials, de Moura is convinced you are up to no good and refuses to allow anybody, even Cook, to set foot in the city, though he will permit you to take on food and water. Annoyed and frustrated, twelve crew members will defy his orders and slip undetected into Rio, where they will be immediately arrested for smuggling and spend a night in a hideous jail cell with other prisoners, who are chained to the walls, only to be released after Cook writes an infuriated letter to the viceroy.

INTO THE PACIFIC

DECEMBER 2, 1768–APRIL 12, 1769

As you sail the 300 miles to Cape Horn, the southernmost tip of Latin America, you will notice the temperature drop dramatically, and Cook will issue your cold-weather clothing. It's a new year, too, and on January 11, 1769, you reach Tiera del Fuego, where you will find locals living in round, beehive-shaped dwellings with wooden frames covered with sealskins and brushwood. They're a hospitable bunch.

After this brief respite, you will face the toughest phase of your trip so far: getting around CAPE HORN. You will have to negotiate the STRAIT LE MAIRE, a funnel-like passage beset by storms that send tidal surges racing through its bottleneck. Three times *Endeavour*, tossed like a twig on the foaming waves, will fail to get through before finally making it out of this deathtrap.

A few days later, you will drop anchor at the BAY OF GOOD SUCCESS, and meet its inhabitants with their distinctive black-and-red body paint, while the botanists head inland to search for specimens. Leaving here on January 21st, there is nothing but the endless horizon of the Pacific in front of you. Time passes slowly. Weeks stretch into months, the monotony somewhat relieved by warmer weather, the variety of marine life on show, and the knowledge—passed on by some of the crew who had been on HMS *Dolphin* when it visited Tahiti in 1767—that your next port of call is home to some of the nicest people on earth.

TAHITI

APRIL 13–JULY 14, 1769

Finally, you will see land, a tiny speck in the ocean called LAGOON ISLAND. You will then pass a few others like it before MATAVIA BAY appears with its palm trees and pristine beaches, the locals coming to greet you in their sixty-foot canoes, bringing a warm welcome, and food to trade for beads and cloth. A decimal system is used for counting; Tahitians have signs for numbers from 1 to 20—enabling them to count up to 20,000—and they follow the lunar calendar and use astronomy to interpret the stars.

TAHITIAN LIFE

At the top of the TAHITIAN SOCIAL SYSTEM is the ARI'I (chief), then the TAHU'A (priests), followed by RA'ATIRI (minor nobility), and MANAHUNE (regular folk). These distinctions aren't immediately obvious, as everyone you meet will appear to be on the same level. There are clues, though, across the island in their MARAE—burial sites. These are often just a pile of rocks, but sometimes quite elaborate memorials made of coral stone with steps, and an altar with carved wooden posts. The largest is at OPOOREONOO—a pyramid with elevent huge steps to a platform. According to legend, the MAUWE, the island's gods, giants with seven heads and superhuman powers, inhabited the earth long before man showed up.

All ranks of Tahitians wear little in the way of CLOTHING, which is made from bark. This is peeled from trees in strips, melded together with fine paste, and beaten flat with large wooden tools, before other strips of beaten bark are added crosswise. The resulting material is dyed in red, brown, and yellow. The Tahitians are keen on TATAUS (tattoos), created using bone and soot, and you might want to get one of their unique tribal symbols as a memento.

There is an abundance of EXOTIC FRUIT AND VEGETABLES available, pigs run wild, and a Tahitian speciality is ROAST DOG. It's worth conquering any aversion you might have to canine cuisine (Cook will record that he's never eaten sweeter meat). First, the dog has all its hair burned off, then its entrails are removed and washed clean. Meanwhile, a fire is started at the bottom of a hole in the ground about a foot deep. When the flames are strong enough, stones are put on top and, once they are red-hot, the fire is extinguished. Green leaves are placed on the

TAHITIAN WOMEN PERFORM THE TIMORODEE. LOOK OUT FOR APPLES.

stones, and the dog and its guts are laid on them with a covering of more leaves, and oven-grilled.

When dining, you will notice that men and women never eat together. This is forbidden (*tapu*). You will also have to bring your own booze if you want any alcohol with your meal; the Tahitians serve nothing but coconut milk and water.

You will be treated to a great deal of MUSIC played on hand drums of hollow wood with a membrane of shark's skin, and four-holed, fifteen-inch-long bamboo flutes, which are inserted in the nostril and blown. The DANCES that these instruments accompany are segregated by gender. The female ones often feature girls on the verge of puberty: during the TIMORODEE DANCE, two groups of near-naked young women split into opposing groups and throw apples at each other. MALE DANCES are more warlike and based around wrestling poses. WRESTLING is popular on the island, so you will get to see the real thing as well.

Property will be the main cause of friction during your stay. The Tahitians have a "what's mine is yours" attitude; private property is a totally alien concept, and "theft," therefore, is a meaningless term. Because theirs is a Stone Age culture, the locals are particularly drawn to objects made from glass or metal. One day after arrival, two of the gentlemen will have their pockets picked, one losing his spyglass, the other his snuff box. From then on, things will keep vanishing. On the whole, they will be recovered without bloodshed, but when a musket is snatched from a sailor's hands, the "thief" is promptly shot dead. After an iron rake goes missing on June 14th, Cook, at his wits' end, will confiscate all the canoes in the bay until it is returned.

There will also be a lot of bartering, done primarily by the Tahitian women, who will exchange sexual favours for coveted objects such as the ship's iron nails. To prevent this, Cook will issue five rules, four of which (the other simply instructs you to be friendly) relate to unauthorised bargaining with the locals; one explicitly forbids trading iron for anything except food. To show how serious he is, Cook will have a seaman flogged for filching nails from the stores.

More often than not, the women will give themselves freely. You will find the Tahitian attitude to sex remarkably liberal. Nor will you witness any resentment, jealousy, or possessiveness from the menfolk; both husbands and fathers seem happy to share. In this relaxed environment, it's no surprise that the majority of the crew will have intimate relations with Tahitian women—some casual encounters, others more serious affairs.

TRANSIT OF VENUS

Regardless of the many delights on offer, your reason for being on Tahiti is to observe the TRANSIT OF VENUS. Cook will order a FORT built to safely store the astronomical instruments—a brass quadrant, a newly patented azimuth compass, a high-quality sextant, and two telescopes with Gregorian reflectors and parabolic mirrors. Work on the fort will start almost immediately on arrival, and you will help construct a high-spiked palisade with swivel guns at each corner, boxing in a series of big tents. For extra security, the two big on-ship guns will be trained on the woods next to the fort.

The quadrant will be set up on May 1st, only to be stolen the following day. Infuriated, Cook will detain all the large Tahitian canoes until it reappears. Thankfully, that evening it is returned undamaged. On the day itself, Saturday, June 3rd, the conditions will initially appear perfect, but then clouds will gather and obscure the view—Cook will complain of "an Atmosphere or dusky-shade round the body of the planet." As a result, none of the three readings taken matches up, making it impossible to get any accurate measurements.

By July 9th, you will be ready to set off again, along with a young Tahitian, TUPIA, who has a smattering of English. However, departure will be delayed by the hunt for two seamen, CLEMENT WEBB and SAM GIBSON, who have not unreasonably decided to desert the ship so they can stay with their Tahitian girlfriends. A search party will be sent ashore and, to ensure the locals' cooperation, their chief is taken hostage. Within twenty-four hours, the two lovesick sailors will be back on the *Endeavour*, and you will begin the next phase of your adventure.

NEW ZEALAND

SEPTEMBER 1769–APRIL 1, 1770

En route to New Zealand, the *Endeavour* will call in at the SOCIETY ISLANDS, whose inhabitants will sacrifice an eighty-pound hog in your honour. It is an isolated highlight. September will bring violent storms, colder weather, and mounting tension; when, if ever, will land appear? Cook knows there is something out there, thanks to the Dutch explorer Abel Tasman, who sighted New Zealand in 1642, but Tasman only skirted the North Island; you will trace every nook and cranny of the whole coastline.

Finally, on October 8th, a panorama of white cliffs, sandy bays, wooded hills, river valleys, and two majestic purple mountains will come into view. As you get nearer, you will see a string of huts on the beach and MAORIS with their warpaint and tattoos

MAORI WAR CANOES—NOT A WELCOMING SIGHT.

paddling towards you in their fifty-foot-long and five-foot-wide war canoes, each capable of holding a hundred men, hurling spears, and making aggressive noises. Your musket fire will force them to turn around, with one warrior killed. An attempt to go ashore fares no better; there is a standoff between you and the Maoris, armed with *patoo patoo*—heavy, hand-held weapons with sharp, serrated blades of green stone. Though Tupia will try to chat with the Maoris and offer gifts, there is more musket fire and another dead warrior. A second landing ends just as badly, with four further Maori casualties. It's no wonder Cook dubs the area POVERTY BAY.

Many of your subsequent encounters with the Maoris will be hostile. At the appropriately named KIDNAPPER'S BAY, Tayeto, Tupia's boy helper, is snatched, and force of arms will be required to get him back, while at the BAY OF ISLANDS you will need a volley from the big ship guns to disperse a force of 200 armed Maoris.

However, there will be friendlier relations, especially at TOLAGA BAY, where you will stay seven days and get a taste of local life and culture. The Maoris are superb fishermen, using lobster traps made from twigs and huge trawling nets. Their clothes are made from a hemp-like plant (*barkeke*), their homes constructed round a frame of sticks and thatched with long grass. You will be pleased to note that every dwelling has a basic outside toilet. Their music is played on wooden flutes and shell trumpets. During their DANCES, the women wear head dresses with black feathers, while the men's WAR DANCES, in their authentic form, will be familiar to any rugby fans.

On November 9th, you will successfully observe the TRANSIT OF MERCURY, before circumnavigating the South Island. You will find a good berth for fishing and supplies of fresh water at MURDERERS BAY near Queen Charlotte Sound, and leave New Zealand on April 1st.

AUSTRALIA (NEW HOLLAND)

APRIL 19–AUGUST 22, 1770

On April 19th comes a historic moment as the lookout spots the coastline of what Cook calls *Terra Australia Incognita*. Searching for a place to drop anchor, Cook passes and names a series of features—Cape Upright, Pigeon House, Cape St George, Long Nose, and Red Point—before landing at a spot that is soon dubbed BOTANY BAY.

As you enter the bay, on April 29th, you will get your first glimpse of ABORIGINES. There will be a group of huts on shore, and two of the inhabitants, one old, one young, will approach the *Endeavour*. Due to mutual incomprehension and suspicion, spears will be thrown and muskets fired, leaving one Aborigine slightly wounded and their huts abandoned. Despite this inauspicious beginning, subsequent contacts with the Aborigines will be peaceable, if tentative. You will notice that the males have beards, bone piercings through their noses, and faces and bodies painted with white pigments. They usually appear armed, carrying three-foot wooden sticks that propel four-pointed bone darts, shields, and boomerangs (be aware that none of Cook's company will know how these work). The women, wearing not much, apart from shell necklaces and bracelets, will keep their distance from you.

Botany Bay is a natural choice for the landing site, as the botanists will have a field day. The wildlife is bewitching and (to your companions) entirely unfamiliar: kangaroos and dingoes, fruit bats, lizards, snakes, huge caterpillars and butterflies, green ants, scorpions, tropical parrots ...

Moving on, you pass PORT JACKSON (Sydney Harbour), POINT DANGER, CAPE MORETON, and CAPE CAPRICORN, before reaching the beautiful but hazardous GREAT BARRIER REEF. On June 11th, it bares its fangs and a stretch of razor-sharp coral will rip a hole in the ship. Cook knows the only hope is to get to shallower water or run aground, so once the sails have been lowered you will immediately set about lightening the load; iron and stone ballast, rotten stores, and all but four of the guns will be thrown overboard.

By the next day, *Endeavour* will be taking on a lot of water and you and every available hand will be manning the pumps. This continuous effort will pay dividends, especially after Cook manages to wrest the ship free and even more water pours in. This will be the crew's finest hour, working together as a well-oiled machine that will prompt the normally sanguine Cook to declare that, "no man ever behaved better than they have done on this occasion."

However, you are still twenty-four miles from shore and still slowly sinking, until one of the crew, Jonathan Monkhouse, comes to the rescue and "fothers" the hole. This entails sewing bits of oakum and wool together into an old sail, then drawing it under the ship to allow the water pressure to force it into the yawning gap in the hull, thereby blocking it and stopping the sea coming in. Job done, you will drift towards shore and come to a halt there on the morning of the 17th.

It will take two weeks to complete the repairs to the *Endeavour*; fixing the hole, scraping barnacles from the hull, and mending the torn sails. You will be engaged in helping to speed this work alon, but there will be time for a little r & r as you camp out on the beach.

By August 5th, you are back at sea, inching your way across the Great Barrier Reef. Its spectacular colors are seductive but the crew never lose sight of the threat lurking beneath the surface: one false move and the reef will tear you up again. There are a hair-raising few hours when the sea falls calm and light winds start easing the *Endeavour* towards certain destruction. Feel free to join the prayers at this point. You will be delivered at the last moment when the wind shifts course and pulls you away from danger.

Having sailed along the whole east coast, you will come to CAPE YORK on August 22nd. Here Cook will name the territory NEW SOUTH WALES and you will stand on deck as he and a few

THE *ENDEAVOUR* UNDERGOING REPAIRS OFF THE EAST COAST OF AUSTRALIA.

men land on the tiny Possession Island, where you will see them raise the British flag, a ceremony marked by three musket shots, answered by three volleys from the *Endeavour*'s guns and three cheers from the rest of you.

HOMEWARD BOUND

AUGUST 23, 1770–JULY 12, 1771

The ailing *Endeavour*—its timber riddled with shipworms, its pumps worn, and its bottom still leaking—will now limp towards the known world, the DUTCH EAST INDIES, and sail past TIMOR, JAVA, and SUVA before dropping anchor alongside Dutch ships and a lone English merchantman in the harbour at BATAVIA.

The Dutch-governed settlement of Batavia has a mixed population—European, Polynesian, and Chinese—of around 20,000 in the city and another 100,000 in the suburbs, and boasts a town hall, churches, and a network of canals, designed to mirror Amsterdam. Unfortunately, after an earthquake blocked the flow of fresh water, the canals have become stagnant pools teeming with millions of mosquitos and caked in foul smelling mud littered with human waste and dead animals. It is an ideal breeding ground for malaria and dysentery, and within days the majority of the *Endeavour*'s crew are on the sick list. (To avoid this calamity, you will be met soon after landing by a company representative disguised as a Dutch merchant sailor. While engaging you in idle chatter, he will slip you anti-malarial medication and pills to ward off dysentery; it is enough to protect you, and you alone, from infection.) By the end of your stay in Batavia, prolonged by the repair work needed to reinvigorate the *Endeavour*, seven crew will be dead from malaria, and forty too sick to perform their duties; Tupia, your Tahitian guide, is the first to succumb.

Back at sea, after a swift stop-off at Prince's Island to take on water, dysentery will raise its ugly head. Over the next few months, another thirty-two of the party will pass away. This will be your darkest hour. Half your shipmates will have perished, along with the two Scandinavian botanists, the remaining artist, and the astronomer.

Everyone's spirits will be lifted a little, though, when the CAPE OF GOOD HOPE is sighted. Two days later, on March 15, 1771, you will arrive in CAPE TOWN. In the idyllic setting of this sleepy outpost, with its whitewashed colonial-style buildings, its vineyards, orchards, and kitchen gardens, and Table Mountain hovering in the background, you can shed some of the stress of the past few months. You will be able to take leisurely strolls through its picturesque streets and visit the local MENAGERIE, with its ostriches, antelopes, and zebras. You will also eat well courtesy of the captain: Cook will purchase a whole ox for you to enjoy.

Somewhat refreshed, the surviving party will gird its loins for the final stretch of the journey. As you CROSS THE ATLANTIC, you will have merchant ships for company, and on May 15th you will witness an ECLIPSE OF THE SUN. When you clear the BAY OF BISCAY on July 7th, you will know the end is in sight. Your first glimpse of England will be on the 10th. By the 12th you will have reached Beachy Head: that afternoon you will drop anchor at the port of DEAL in Kent.

DEPARTURE

It is important not to get caught up in the riotous celebrations that will greet the *Endeavour*'s return. Deal is a small but bustling port with a reputation for hard drinking, brawling, and vice in all its forms. It is also a favoured hunting ground for the Impress Service; you could easily find yourself press-ganged again and on a ship going god knows where. Instead, find the Dover Road and walk south for about two miles until you reach the village of UPPER WALMER. There you can safely stop at the ROYAL STANDARD INN for a quick drink to toast your survival before strolling a short distance to OLD ST MARY'S CHURCH, from where you will DEPART.

PART FIVE

Extreme Events

The Eruption of Vesuvius

AUGUST 23–25, AD 79 ✵ POMPEII

THE ERUPTION OF MOUNT VESUVIUS WAS not the largest or deadliest in history. Yet it holds a special place in our imagination thanks to the ruins it left behind. Year after year, thousands troop around the excavations at Pompeii and Herculaneum hoping to recapture something of their Roman life, so movingly "caught in time." Now you can experience it in *real time*. This extreme travel trip gives you the chance to sample the day-to-day life of these energetic and cosmopolitan Roman towns—as well as to witness an awesome and spectacular act of nature. Having done so, you will be in a position to fully appreciate the scale of the tragedy as it unfolds before your eyes. At some point during your stay, you may become overwhelmed by the impulse to suggest to some of the locals that they might want to leave town for a few days. And take all their possessions with them. This is strictly forbidden. If you feel you are unable to control your charitable, humane instincts, then

Vesuvius is not the trip for you. Interaction with the citizens must not become too intimate or friendly.

You will spend a day and night in POMPEII and a day in the seaside town of HERCULANEUM, before boarding a specially chartered SHIP in the Bay of Naples, from where you will watch the cataclysm. The pitch-black clouds of ash and debris produced by the eruption cloak the surrounding countryside for miles in all directions, making it impossible to find a decent vantage point on land. At sea, however, you will get an unobstructed, if long-distance, view of the ghastly proceedings.

BRIEFING: VESUVIUS

POMPEII's extremely pleasant climate, proximity to the sea, and fertile soil has meant there has been a settlement here since the sixth century BC. The city came under the control of Rome somewhat reluctantly in the fourth century BC and joined a failed rebellion against its masters in 89 BC. After that, however, it was integrated peacefully into the empire and has thrived; its population is currently around 12,000, with another 24,000 in its rural area. They are sustained by some of the best agricultural soil in Italy—the mineral-rich volcanic earth produces an abundance of cereals, grapes, olives, apricots, figs, peaches, almonds, and walnuts, as well as being ideal for grazing sheep. This plethora of goods is traded in the city and exported across the empire, while imports of more exotic fare—spices, perfumes, cloths, jewellery—flow in, making Pompeii an important centre for merchants.

HERCULANEUM shares a similar history to Pompeii, though it is a little smaller, and wealthier. It has some of the empire's finest villas in the town and vicinity.

VESUVIUS, the author of both cities' destruction, is a so-called "humpbacked" mountain, formed by a collision of two tectonic plates, the African and Eurasian, roughly 25,000 years ago. Before AD 79, there have been at least three other major eruptions, but none matched the ferocity of what you are about to witness, in which Vesuvius spewed out molten rock and pulverised pumice at a rate of 1.5 million tons per second, forming hydrothermal pyroclastic flows—molten lava with an accumulated thermal energy 100,000 times greater than the atomic bomb that levelled Hiroshima.

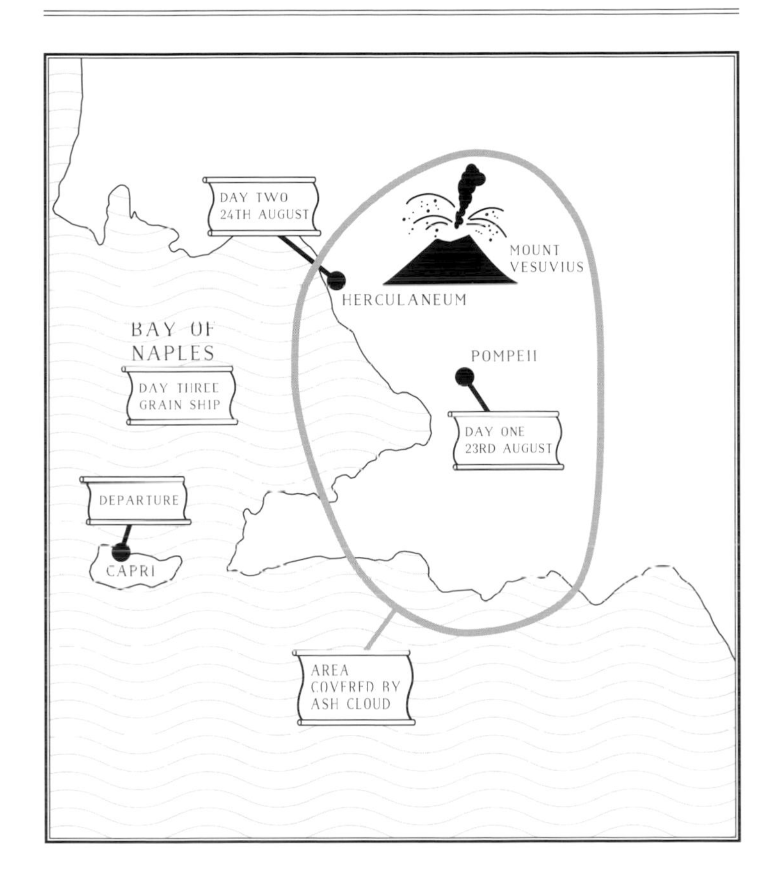

THE TRIP

You will arrive just outside the city of POMPEII, near family burial grounds marked by memorials of all kinds, ranging from large, ornate, multistorey structures honouring the rich and powerful, to the simplest of gravestones. You will be wearing a brightly colored toga to help you blend into your environment; Pompeii is a city that likes its colors, and the houses are painted in bright reds, yellows, and blues. After a short walk, you will enter the city via the NORTHEASTERN GATE.

ROOMS, FOOD, AND NIGHTLIFE

You will be staying in exclusive RENTED APARTMENTS on the upper floors of multipurpose buildings near the centre of Pompeii. These light, airy, and spacious condos, with large windows and terraces, luxury fixtures and fittings, and their own TOILET (a wooden seat over a chute carved in the wall, emptying into a cesspit down below), will give you the chance to relax and unwind after the day's exertions. A small potable oven (*farnus*), with its own grill, provides a self-catering option.

There are, however, plenty of eating options around the city, and the fresh FISH DISHES are highly recommended. Do take the opportunity to sample *garum*, the fish sauce that is something of a speciality. Made from a mixture of sea salt and seafood, fermented in a vat for a few months in the sun, then decanted into jugs, the very best comes from pure mackerel (*liquaminis flos*). But be wary: the low-grade stuff is no better than eating rotten seafood.

For carnivores, the most-often-served meat is PORK, which usually comes in the form of sausages or black pudding. For the unsqueamish, there is the Roman delicacy of DORMICE. Fed on nuts, the dormice are fattened up in specially designed jars, before being stuffed with pork, peppers, and nuts, all glued together by *garum*, and cooked until tender. Some of the best-quality produce can be found in taverns around the MACELLUM (meat market) situated in the FORUM.

Another option is STREET FOOD or buying directly from retailers. There are BAKERS everywhere supplying fresh bread from their premises, using big ovens, not unlike present-day pizza ones. The dough is prepared in an adjacent room, while out back is the flour mill powered by a beast of burden. A loaf of bread with some local cheese and olive oil may be all you need for lunch. Don't hesitate to sample the LOCAL WINES. They have a decent reputation, particularly the Falernian brand.

You may want to take your evening meal early, since if you venture out after dark you'll need to keep your wits about you. With no light except the stars above, and all the houses, shops, and workshops shuttered up, the streets are monopolised by drunken revellers and muggers.

If you do want to take in some local nightlife, though, you can take your pick from plenty of TAVERNAS. Aside from drinking, these play host to dice-based board games. GAMBLING is the norm, and if you want to avoid trouble it's best to watch rather than participate. Note also that these tavernas are exceptionally male environments. Female travelers may find the ambience intimidating and will almost certainly attract unwanted attention, as the assumption will be that you are a woman of easy virtue.

AUGUST 23RD: POMPEII STREET LIFE

POMPEII is not a large city, and the best way to explore it is on foot. Wheeled transport is a nightmare, the narrow, single-lane roads choked with traffic as traders move their wares in carts drawn by donkeys or mules, leaving no room to maneuver. Congestion is such a problem that you will see traffic-calming measures all over the place, including signage and one-way streets.

Even crossing the road can be a hazardous affair, due to the garbage and waste, both human and animal, scattered in your path. Thankfully there are high, sturdy pavements and, to avoid treading in anything untoward, there are stepping stones from one side of the street to another, with gaps between for cart wheels to pass through. It's going to get hot in the August sunshine, so remember to keep hydrated. Luckily the locals have that covered: at regular intervals, often at major junctions or crossroads, are WATER FOUNTAINS with large spouts and continuously running water supplied by a tank underneath. The city's water is entirely safe to drink. It comes from an aqueduct that sends it down the mountain to a tall water castle just outside the walls. It then distributes the water to a dozen or so towers made of stone and brick, up to twenty feet tall, with a lead tank at the top, which control and contain the flow of water through pipes below the pavements.

Pompeii is full of vivid, vibrant imagery everywhere you turn. There are large formal SCULPTURES of prominent citizens and figures from history, and shrines dedicated to the ever-present gods. Look more closely and you will also see dozens of smaller figures carved into the street or next to shops and businesses. Particularly common are dwarf-like figures with huge PHALLUSES; aside from being a symbol of fertility and male power, the phallus also represents good luck.

Most striking is the GRAFFITI. The life of the city is recorded in these scrawls, etched into every available surface. The walls will speak to you of politics and public affairs. They advertise forthcoming events and honour local celebrities. They boast of sexual

VESUVIUS DEPICTED IN A FRESCO OF BACCHUS AT POMPEII. THE MOUNTAIN WILL NEVER LOOK LIKE THIS AGAIN, AS ITS ENTIRE CONE WILL BE BLOWN OFF BY THE ERUPTION.

conquest and desire, of love gained and lost, and feature quotes from the great poets and writers of antiquity.

This tradition of popular artistic expression extends to the SIGNS outside shops. The best feature intricate, detailed images, such as a builder's with symbols of his trade—tools, chisels, mallet, and the ubiquitous phallus. The sheer volume and diversity of retailers and craftsmen—textile merchants, fashion houses, barbers, perfumers, jewellers, cobblers, spice stalls, and so on—makes for excellent window shopping. You will probably notice a number of brothels (*lupanari*), too. Pompeii has quite a reputation in this respect, and Roman men will pop in during the day for a service.

As you wander round, you will notice that large parts of Pompeii are undergoing RECONSTRUCTION AND RENOVATION, with many of its most prominent public buildings out of commission. In some cases the rebuilding work is well advanced, in others there is little to see except rubble. For example, the AMPHITHEATER, site of gruelling gladiatorial contests and blood-soaked spectacles involving wild animals, is closed due to construction work. This disruption is the result of a major earthquake that struck the city on February 5, 62 AD, measuring 5 to 6 on the Richter scale. On the day of the quake, the Forum had been packed with people attending two sacrifices to mark the anniversary of Emperor Augustus taking power and a feast to honour city gods, adding to the chaos and the death toll.

Entering the FORUM, a great open space in the heart of the city, the impact of the earthquake will be evident. Flanked by colonnades on all sides, it is the home of two major temples: the TEMPLE OF APOLLO, the current version dating back to the second century BC, and the TEMPLE OF JUPITER, JUNO, AND MINERVA, which was severely hit by the quake. There is also a house dedicated to the priestess EUMACHIA, daughter of a well-to-do wine merchant, plus the BASILICA, the largest public building in Pompeii, the surface of which is covered in graffiti. Resist the temptation to add your own "I was here" to the hundreds of messages already carved into its façade.

The TEMPLE OF APOLLO, with its exterior columns and triangular gables, is well worth a look. Climb the steps up to a podium and then go in through its high doors and you will be greeted by statues of the relevant gods and goddesses, alongside dedications and tributes left by worshippers. Otherwise, the interior is relatively austere, a plain, simple space, not designed for major religious ceremonies or rituals (which take place in the Forum proper).

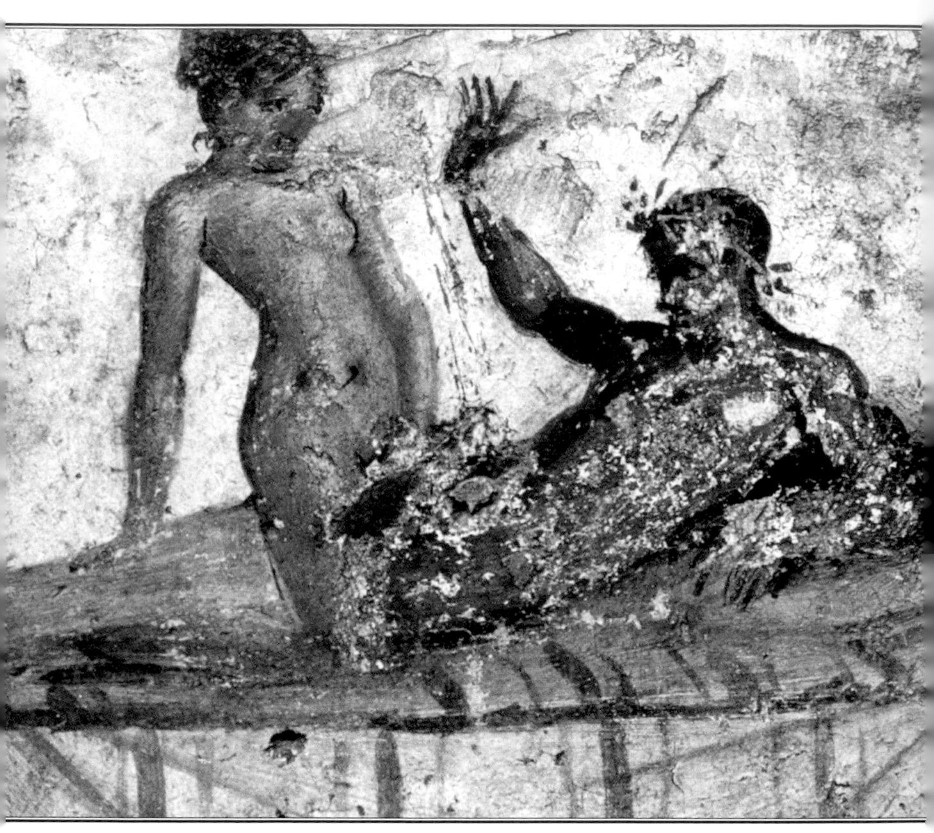

UP, POMPEII! ROMANS HAVE A RELAXED ATTITUDE TO SEX, AND YOU'LL SEE A RICH VARIETY OF ACTS DEPICTED IN BOTH PUBLIC AND PRIVATE BUILDINGS.

No day in Pompeii would be complete without a visit to the BATHS. You will need to be comfortable with nudity, as everybody will be naked, though currently the facilities are segregated by sex. Travelers hoping to indulge in mixed bathing will be disappointed that the Central Baths, the only venue in the city where men and women mingle together, is still undergoing repair work from the earthquake.

Once inside the baths—the front entrance is for men, a side entrance for women—you ditch your toga in the changing rooms (*apodyterium*). This is worth a close look, as, to help you find where you have left your clothing, there are a series of erotic murals featuring depictions of unashamed cunnilingus and even a threesome involving two men and one woman. These are part of the Roman acceptance of the erotic, and not given more than a passing look by your fellow bathers.

Toga off, you are free to exercise, swim, sunbathe, have your skin scraped, or get a rubdown. The main attractions are the various rooms offering a full range of temperatures to immerse yourself in: a cold room (*frigidarium*), a warm room (*tepidarium*), a hot room (*caldarium*), and, for the hardcore enthusiast, the hot sweat room (*laconium*). The one thing we'd not advise is a dip in the pool, which is a breeding ground for bacteria of all kinds.

At some point in the day you may well need to relieve yourself. There are a number of PUBLIC TOILET facilities available, but don't expect the privacy of your own cubicle; emptying your bowels is a communal activity in Pompeii. You will squat in a row next to your neighbor on a keyhole-shaped seat cut out of a length of marble slab, with a trench directly underneath to collect and flush away the waste, and a small trough in the floor in front of you filled with fresh water. Once you're done, take a SPONGIA (a sea sponge attached to a stick), dip it in the trough, slip it discreetly under your toga, and clean up as necessary.

AUGUST 24TH: HERCULANEUM

You will rendezvous early next morning at the arrival point, then head to the nearby town of HERCULANEUM. Small, with a high percentage of wealthy citizens occupying large, opulent homes, Herculaneum's position by the sea makes it an ideal spot for holiday homes for the Roman elite. It is also the perfect spot to view the FIRST ERUPTION.

At 1pm on August 24th, VESUVIUS bursts into life, propelling a thick cloud of ash and stones thousands of yards skywards. While this thick cloud cloaks Pompeii, Herculaneum, which is to the west of Vesuvius, only gets a very fine dusting, leaving a sooty layer a few inches deep. Though you might be troubled by irritated eyes, a dry throat, tight chest, and nasal congestion, you will get a clear view of the action over Pompeii as the blanket of foul ash descends on it.

Though some of Herculaneum's residents stay put, hoping the worst is over, most abandon their homes and hit the road. With many fine dwellings abandoned, this is an excellent opportunity to explore.

Once you have selected a suitably DESERTED HOUSE, you will find yourself entering an open courtyard, the ATRIUM, from which you will be able to access the maze of rooms. There will be much to admire: shrines to the household gods (*lares*), ornate mirrors, oil lamps, and bronze sculptures, and FRESCOES covering walls, ceilings, and even floors, depicting scenes from mythology, erotic exploits, and hunting expeditions. Domestic GARDENS are a source of pride for their owners. Some will be wild, overrun with trees and vines; others more formal, with neatly arranged flowerbeds, or ornamental in style, featuring fishponds, shimmering pools, and exquisite shrines. As you investigate, you will be struck by the eerie silence and ghostly hush that has descended on these homes, where only inanimate relics—the unoccupied dining room couches, the half-eaten food, the clothes dumped carelessly—bear witness to lives suddenly interrupted. They will remain in this state of suspended animation for a few more hours until the deluge consumes them.

By NIGHTFALL, you should be heading for the shoreline, where small rowing boats will be waiting to take you out into the Bay of Naples, where you will transfer onto a larger vessel. The last shuttle will wait until 11:30pm, not a moment later. Anybody who fails to make the rendezvous in time will be stranded with no way back to the present. It is not a good prospect. Herculaneum will be flattened by hot gases sweeping through it at 100 miles per hour, reaching a temperature of 932°F (500°C), then buried under sixty feet of ash.

AUGUST 24TH: BAY OF NAPLES

Your rowing boat will help you transfer to our charter, a GRAIN SHIP with two sails, capable of an average speed of 4-5 knots in a favourable wind. It is controlled by side rudders and steering oars located at the stern of the ship; these are manipulated by a system of cables rotated on an axis which are operated by the steersman who works the tiller (*clavus*), a bar set perpendicular to the oar. The steersman and the rest of the crew are all highly trained company employees. Light refreshments—soft-boiled eggs in pine-nut sauce, lentils with coriander, fresh oysters, sliced roast boar—will be served.

Gathered on deck, you will be ready for the main event later that night. When the initial ash cloud produced by VESUVIUS collapses, the gases it releases expand to create a PYROCLASTIC SURGE. This process will keep repeating itself in a series of devastating pulses, generating temperatures of up to 572°F (300°C). The third and fourth waves will level POMPEII, either incinerating or suffocating every living thing in their path.

From the boat you will see the top of Vesuvius smouldering away, burning brightly in darkness so thick it blots out the heavens as night merges seamlessly into day. Dotted across the landscape, the intense heat produces spots of glowing light like hundreds of bonfires.

You should also be able to make out the frantic activity on the shoreline as desperate throngs of people try to escape, trapped between increasingly choppy waters and the encroaching eruption. Many perish under hails of flaming debris—great masses of rocks and stones that have been transformed into lethal white-hot coals. As the catastrophe unfolds on land, you will notice the sea getting decidedly rougher, with huge, rolling waves slamming into the boat, tossing it hither and thither. This is because of seismic activity; the eruptions create tremors that ripple through the seabed.

Even on your boat, this is not a trip for the cautious. Seasickness may be an issue, and you can expect slippery decks, an occasional soaking, and moments when it will seem as if the boat is about to sink without trace. Try to stay calm. It is a sturdy vessel and should be able to cope with the extreme conditions. We haven't lost one yet.

Shortly after the sixth and final pyroclastic surge, the ship will sail to the island of CAPRI and deposit you on a beach, ready for departure.

The Peasants' Revolt

JUNE 12–15, 1381

MERRY OLD ENGLAND HAS REACHED ITS boiling point. King Richard II, just fourteen, is perceived to be in thrall to a coterie of traitors and evil advisors: John of Gaunt, Duke of Lancaster, and effectively regent during his nephew's minority, Archbishop of Canterbury Simon of Sudbury, and the royal treasurer Robert Hales. After three decades of social upheaval and high taxation, join the peasants, yeomen, and free townspeople of Kent, Essex, and London as they rise up against their oppressors. You will join the rebel bands at Blackheath, consort with its leaders WAT TYLER, JACK STRAW, and the radical preacher JOHN BALL; meet the KING'S BARGE AT ROTHERHITHE and then, with the angry London mob, STORM LONDON BRIDGE and take the city on the feast of Corpus Christi. This is an unparalleled opportunity to experience medieval London, its architectural eccentricities, rich social mix, and odours, but don't linger for too long; all across the city you can see and join the febrile arc of rebellion, mob anarchy, and counterrevolution.

BRIEFING: ENGLAND IN REVOLT

The PEASANTS' REVOLT emerged from three decades of tumult in medieval England. The Black Death that had swept Eurasia in the 1340s and 1350s killed over one-third of the population of England. With land plentiful and labour in short supply, the fundamental balance of power between the nobility and the peasantry shifted in favour of the latter. Bondsmen evaded their obligations, and moved to other manors or towns, and the cost of labour soared despite repeated attempts to legally control wages. Serfdom was collapsing despite the draconian ORDINANCE OF LABOURERS and STATUTE OF LABOURERS, which tried to keep them in place.

A widespread breakdown in feudal authority was made worse by the death of Edward III in 1377, when his son, now Richard II, was only ten; rule by his regent, his uncle John of Gaunt, the Duke of Lancaster, was widely reviled. The disastrous foreign policy conducted by Gaunt (who will be in the north of England during the rebellion signing another peace deal with the Scots)—above all, the long-running and unwinnable Hundred Years' War with the French—required the levying of three monumentally costly POLL TAXES on the whole English population in 1377, 1379, and 1381. The first two had been met by some resistance and much evasion, but it was the third poll tax, due to be collected in June 1381, that finally ignited widespread protests in Essex and Kent. These led to systematic attacks on tax collectors and landowners and the organised BURNING OF LEGAL RECORDS of estates, churches, and religious orders that specified feudal obligations and rents.

The rebel movement positioned itself politically by claiming to be for "KING RICHARD AND THE COMMONS," and only against the traitors and incompetents who had led him astray. So when the message came to the Kentish rebels that the king would receive them near Blackheath, then outside London, they marched on the capital in their tens of thousands.

THE TRIP

Visitors will be arriving around lunchtime on Tuesday, June 12, 1381, on the scrubby open lands of BLACKHEATH. These lie on the south side of the Thames about a mile upriver from the Tower of London and the city itself. You will find yourself among a

THE PRIEST JOHN BALL (ON HORSEBACK) ENCOURAGING WAT TYLER'S REBELS—DEPICTED, ODDLY, IN FULL ARMOUR IN THIS 1470 MANUSCRIPT OF *JEAN FROISSART'S CHRONICLES*.

gathering band of provincial rebels and lower-class Londoners, awaiting a meeting with King Richard II.

For our more intrepid medievalists, we are now making available two longer excursions. These will allow you to experience the initial risings in Essex and Kent before arriving at Blackheath; though note that a basic level of equestrian skills and physical fitness will be required.

ESSEX, ARISE

A full twelve days longer than the London-only excursion, this trip puts you down on May 30th at the edge of the market town of Brentwood, Essex, where a meeting of the local Justices of the Peace with villagers will erupt into organised protest, the crowd refusing to pay their taxes before they run the king's representatives out of town. A short ride on WHITSUN SUNDAY to the village of BOCKING will allow you to join a great gathering of Essex men and witness the explosion of discontent and its organised export across the county.

Over the next week, the southern half of Essex, all of it within a day's ride, offers wildfire action of all kinds. Visitors will be expected to join the gathering of Essex rebels on JUNE 10TH at CRESSING TEMPLE near the village of Coggeshall, where the Sheriff of Essex will be killed that evening by rebel bands. The MARCH ON LONDON begins on JUNE 11TH via the town of CHELMSFORD, where there will be a spectacular display of RECORD BURNING. Please note that this march will terminate at Aldgate or Ludgate, on the north bank of the Thames. You will need to make your own way south across London Bridge to Blackheath, though you may find that the gates to the city are firmly shut. You should, however, be able to catch a ferryman on the wharves east of the Tower of London.

KENT IN FLAMES

Visitors will arrive in DARTFORD, the epicentre of the revolt in Kent, on JUNE 5TH. Rebels have been riding out to the surrounding villages to gather support, and you should find a couple of thousand armed men mustering there. On the following day, JUNE 6TH, you will join them on their short march to ROCHESTER, where they will take the thinly defended CASTLE and take hostage the constable of the castle, Sir Roger Newenton. The 7th takes you to MAIDSTONE, where the mob will be STORMING THE JAIL and releasing the charismatic and heretical preacher of liberty JOHN BALL, whose Rotherhithe sermon later in the week is not to be missed.

The most energetic visitors will then be able to join the emerging rebel leader WAT TYLER, whose bands will be marching on CANTERBURY some thirty miles to the south. On JUNE 10TH you will be able to join Tyler on his entrance to CANTERBURY CATHEDRAL, where he will call for the election of a new archbishop, given the corruption of the old one. He will also call for the townspeople to name any "traitors" in their midst. Three beheadings and a town-wide riot will follow while the rebels take the castle and break open the jail. On the following morning Tyler will recieve a royal messenger with the proposal that the rebels meet KING RICHARD in person near BLACKHEATH. For those who would like to take a mid-revolt breather, we recommend waiting things out in MAIDSTONE or DARTFORD (Rochester is particularly prone to fire and riot), from where you will be able to rejoin Tyler's returning band as it heads towards Blackheath in the late afternoon of June 11th.

Whichever package you have chosen, and whatever route has brought you to BLACKHEATH, take a moment now to orient yourself; you will need to return here for your departure on the morning of June 16th. The small hamlet of GREENWICH is immediately to your north and it is here, on the low hill, that most of the leadership of the rebellion will be gathering. Follow the track across the heath away from the river and you will see two right-hand turns. The first will take you down to ROTHERHITHE, where the rebels will be meeting King Richard; the second takes you to SOUTHWARK, the notorious suburb that sits on the south side of the river opposite London Bridge.

CITY LIFE

There are 354 TAVERNS within the walls of the city of London, so getting a drink shouldn't be a problem. Note, taverns are the upper end of the market serving only wine and food. The presence of linen tablecloths is a good guide to the best of them. The standard tipple is a BORDEAUX RED at a penny a cup; don't ask for claret, as the term won't come into use for another couple of centuries. Alternatively, LEPE and OSEY are decent and strong Iberian whites. Better, but more expensive, RHENISH WINES at two pence a cup should also be available. Food in these establishments will revolve around ROAST MEATS and PIES.

London also boasts innumerable ALE HOUSES, with an altogether rougher clientele and more raucous atmosphere. Do keep an eye on your purse and your back in these establishments. As well as ALE, CIDER and MEAD should be on sale, and basic nibbles like DARK BREAD and CHEESE. Street-food sellers are normally a big feature of fourteenth-century London, but the state of chaos that will prevail for most of your visit means that they will be around less than normal. Nonetheless, do look out for enterprising PIEMEN, BAKERS, and FRUIT SELLERS. A word of warning on MEAT PIES: unscrupulous bakers have been known to mince up rotting carcasses to fill their wares.

London food markets and merchants would normally be a healthier alternative: in the month of June they should be full of soft fruits and spring vegetables. But, again, conditions of urban riot and mob rule will make trading difficult over the Corpus Christi weekend.

WHAT TO WEAR

The changing class structure of fourteenth-century England, and the threat to the established order of social mobility from below, has seen the introduction of increasingly punitive sumptuary laws. These have specified the kinds of fur, emblems, cloth, and types of clothing that each layer of the social hierarchy can wear. But, like the rest of the feudal law, it has been breaking down. Over the main weekend of the Peasants' Revolt, for security reasons it will be best to dress modestly. It would be unfortunate to be mistaken for a lawyer, tax collector, or juror during the worst of the mob violence in London. For both men and women travelers, we suggest clothes befitting someone above the rank of yeoman, perhaps a lowly townsperson or a junior scribe. We recommend: for gentlemen, HOSE (but not brightly colored), a THIGH-LENGTH COAT, and DOUBLET (go easy on the fur on the cuffs and the collars); for ladies, a LINEN SMOCK, a plain GROUND-LENGTH TUNIC, or KIRTLE, with narrow sleeves and surcoat.

ACCOMMODATION

Advance reservations have been made for you in two locations: outside of the city walls at the TABARD INN in Southwark and within the city walls at the LAMB TAVERN in Leadenhall Market. The former will be a safer and more secure evening venue, but the Lamb Tavern does keep you very close to the action. Feel free to use both. Payment will be required in advance and on arrival. Normal prices would be in the region of 1 pence for a bed for the night, 2 pence for a meal (more for meat and wine). We suggest budgeting for 4 pence per person at least. Your HORSES will need stabling and fodder, and that's another 2 pence each. Given the breakdown in social order, be prepared for a certain degree of price inflation. Beds are basic wooden frames strung with rope, but you can expect at least two straw-and-hemp mattresses per person.

HEALTH CARE

The last major outbreak of bubonic plague in southern England was in 1369; nonetheless, all travelers will need to be fully up to date with PLAGUE VACCINE. In addition, visitors must be inoculated against TYPHOID, TUBERCULOSIS, and HEPATITIS. However, the most likely medical problem you will encounter will be in your guts and bowel. Emergency supplies of imodium will be available to travelers.

On such a short trip the danger of contracting LEPROSY is very slight; so do not be alarmed by the presence of lepers in London and on the roads in Essex and Kent.

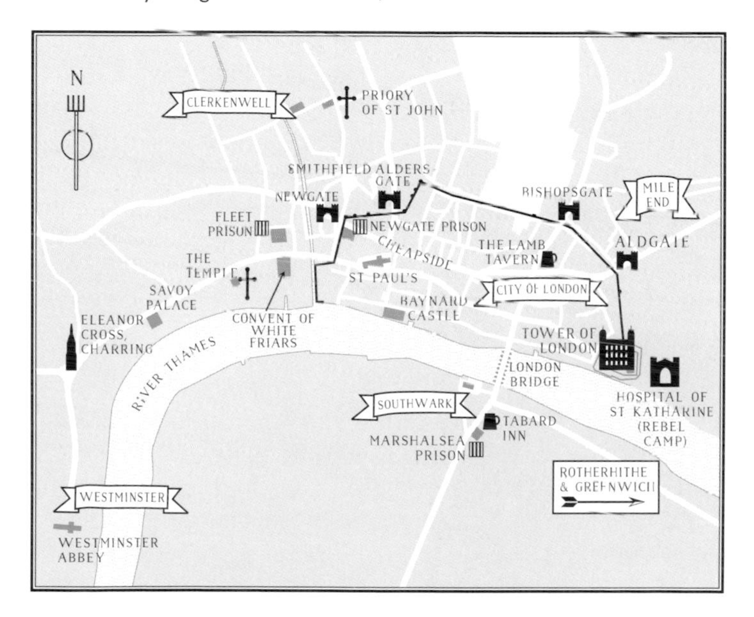

TUESDAY, JUNE 12TH: MEETING THE REBELS

The crowd on BLACKHEATH will be tired, hungry, and expectant. Much of the day will be consumed by gathering firewood for the night ahead, pitching what little cover people have brought with them, and waiting for a message from the king. Listen out for the FANTASTICAL RUMOURS that are sweeping through this volatile temporary encampment; Queen Joan apparently has blessed the rebellion, and the Earl of Buckinghamshire is about to declare for the people. The rebel leadership, on the hill near the river, will send their hostage, SIR JOHN NEWENTON, to the Tower of

London just a mile or so away. In the late afternoon, you may see two mounted royal heralds approach the hill. They will be delivering a summons from the king to meet at Rotherhithe the next morning.

An interesting afternoon and comfortable evening are available in SOUTHWARK (where you have a room booked at the Tabard Inn). We suggest that you arrive there by 4pm at the latest, as by 6pm much of the neighborhood will be in flames. The first target of the rebels, whom you will see joined by many Southwark townsmen, will be the MARSHALSEA PRISON next door to the inn.

RICHARD II ALMOST MEETS THE REBELS AT BLACKHEATH. THE TOWER OF LONDON IS DEPICTED BEHIND THE ROYAL BARGE.

Over the course of the evening, a number of houses belonging to Richard Imworth, keeper of the king's Bench Prison at Southwark and a local hate figure, will be burned, as will the houses and records of identified jurors and tax collectors. In the late evening many rebels will head west out of Southwark and march on LAMBETH PALACE, the home of the Archbishop of Canterbury. You should find this mob relatively restrained. The main purposes of the occupation will be finding and burning archiepiscopal records and the opening of the bishop's casks of wine. The party going on at the palace kitchen is particularly lively; listen out for the popular song of the day, "A Revel and a Revel."

WEDNESDAY, JUNE 13TH: TYLER, THE KING, AND AN EVENING OF BONFIRES

Whether you spend the night at Blackheath or bed down in Southwark you will need to be up early this morning to JOIN WAT TYLER and around a tenth of the rebel army on the south bank of the Thames at the royal manor of Rotherhithe. Tyler should be near the two large St George flags flying above the crowd. While you are waiting for King Richard and his entourage to put in a showing, the highlight of the morning will be the sermon given by JOHN BALL, the radical preacher freed by the Kentish mob in Rochester on June 6th. Listen out for his use of the well-known proverb "When Adam Delved and Eve Span, who then was the Gentleman?" It's the kind of simple egalitarianism that sends shivers down the spine of any feudal ruling class.

Keep your eyes to the left. KING RICHARD'S BARGE and four accompanying vessels will appear around the long curve of the Thames. The king will be accompanied by ARCHBISHOP SUDBURY, Treasurer SIR ROBERT HALES, and the EARLS OF WARWICK AND SALISBURY. Expect a lot of baying and ballyhoo onshore until the royal fleet pulls up about thirty yards away and refuses to land. There will be some discussion across the water and then a flurry of activity as the rebel leaders respond to the king's request for a

WRITTEN PETITION. One man will walk out into the water with the document and hand it over. It is a long list of required reforms and specific calls for the execution of leading members of Richard's court like John of Gaunt and Bishop Courtney of London, as well as Archbishop Sudbury and Treasurer Hales, who are on the boat. After some discussion, the royal retinue will turn around and row off upriver, leaving the Kentish rebels on the shore in a state of uproar.

From late morning onwards, the now-angry rebels will be marching from Rotherhithe and Blackheath towards Southwark and London Bridge. Look out for the BROTHEL by SOUTHWARK FISHPONDS, owned by William Walworth, mayor of London, which will go up in flames as the rebel army passes by. Many rebels will then gather on the southern third of LONDON BRIDGE, their path blocked by its raised drawbridge. On the other side, past the shops and the chapel that line the stone bridge, a large crowd can be seen, many of whom will appear to be welcoming the arrival of the rebels. At this point it is is widely believed that they are not a mob, and that they have come only to serve notice on traitors and have promised to pay market prices for victuals in the city. It won't quite work out that way. The drawbridge will eventually be lowered and, to cacophonic hooting and wailing, the crowd will pour across it and into the city.

Once you are across and heading north on BRIDGE STREET, take a moment to pause at the first junction; BILLINGSGATE is on your right and THE ROPERY is on your left. Some of the crowd streaming around you will turn right down Billingsgate and head for the TOWER OF LONDON, where a large crowd will be gathering on Tower Hill taunting the royal retinue trapped inside. Some will head straight on towards Fenchurch Street and then to ALDGATE, where the massed ranks of many of the Essex rebels are waiting to enter the city. While in this part of town, do look out for GEOFFREY CHAUCER, who is currently renting an apartment above Aldgate. Most of the crowd will be turning left and heading west towards Ludgate and Newgate, where once again many other rebels are encamped. The NEWGATE MOB will be making for two locations on the other side of the city walls: NEWGATE PRISON and, three-quarters of a mile north up Golden Lane, the PRIORY OF ST MARY in Clerkenwell, the headquarters of the Knights Hospitaller.

By far the largest mob will be heading west towards Ludgate; follow them and keep the spire of St Paul's Cathedral to your right. LUDGATE itself is worth a look. It has recently been rebuilt using stones taken from the houses of rich Jewish merchants (killed or expelled in the thirteenth century) and is adorned with statues of ancient Britannic monarchs—including the legendary King Lud. Beyond the gate the city will give way to the pastures, estates, orchards, and grand manors of the Bishop of Salisbury and the CARMELITES, or the White Friars, as they are known. Very few rebels will enter these properties, though a small detachment will turn north to break down the doors of FLEET PRISON. Nearly everyone else will be heading along Fleet Street to either THE TEMPLE or, another mile down the Strand, to the largest house of all—John of Gaunt's SAVOY PALACE.

The Temple had originally served as the headquarters of the Knights Templar in the twelfth century, but after their dissolution has fallen into the hands of the Knights Hospitaller. They in turn had then leased the whole place out to the London legal profession, providing lodgings for trainee lawyers and repositories for legal documents. Around 4pm the grounds will be stormed by the rebels, some of whom will actually dismantle the outbuildings, while a gigantic BONFIRE of rolls, remittances, books, and case notes will go up in flames in the gardens.

Attention will then turn to the Strand, the continuation of Fleet Street heading west. Here you will find the great houses of the Bishops of Exeter, Bath, Llandaff, Coventry, and Worcester, but more opulent than any of these is the SAVOY PALACE standing on the south side of the Strand about halfway between the city and Westminster. There will most likely be a considerable crowd here by the time you arrive, and the large gates in the palace's formidable stone walls will already be off their hinges. Still true to their original intent of punishing but not looting, the sack of the Savoy Palace is strangely ordered. Its napery, clothes, and cloth will be burned to a cinder. Jewels will be ground to dust rather than being pocketed; gilt plate, both silver and gold, will be beaten out of shape rather than stolen. One attempt, mid-evening, by a rebel to pilfer some of the palace's plate will result in an impromptu court and punishment. Please don't be tempted inside the building by the sounds of the drunk laughing or the righteous calling; the small fires already lit inside the palace will be turning into a raging conflagration when three barrels of gunpowder inside are accidentally ignited.

EVENING

Towards evening, the action turns to CHEAPSIDE, which should be approached with some caution. This rare public space, right in the heart of the city, has long been a place of preaching, public punishments, and general milling about. It sits at the intersection of Milk Street, Bread Street, and Wood Street, and can be easily identified by its ELEANOR CROSS, a structure made of four hexagonal stone steps crowned by six statues of the late Queen Eleanor (it was built by Edward I, mourning the journey of his dead wife's body from Lincoln to Westminster Abbey). This evening, and for the next few days, it is going to be a site of EXECUTIONS by clumsy beheadings. The mob will be looking for a variety of known tax collectors and questermongers all over the city, and in particular the notorious ROGER LEGETT. Legett will be cornered by the mob at the church of St Martins le Grand on Newgate Street, a traditional place of sanctuary. The mob, however, will be having none of it. Legett will be dragged out of the church, and down West Cheap to Cheapside and his doom.

Alternatively, you can join the crowds heading north along the Holborn river towards the open fields of CLERKENWELL. En route, the mob will be burning a series of properties owned by Legett, but the main prize is the PRIORY OF ST JOHN OF JERUSALEM—the home of the Knights Hospitaller, a fabulously wealthy military order. The entire complex will be torched, and the remains will burn for over a week.

If you need to get away from the claustrophobia of the city streets or the mania of large-scale arson, head west towards the TOWER OF LONDON. In addition to the gathering on Tower Hill, you will now find a REBEL VILLAGE forming on the fields of the HOSPITAL OF ST KATHARINE on the far-eastern side of the Tower. Late in the evening two ROYAL KNIGHTS will arrive at St Katharine's with a parchment bearing the king's own seal. One man among the crowd will stand on a hastily provided chair and read out their message, which is to the effect that "everyone will be pardoned if they'd just go home now and then all put their complaints in writing." This will elicit a very strong reaction from the overwhelmingly illiterate crowd, who have been burning written records for the past week. Some of them will set off back to the city this evening to burn the houses and papers of lawyers and recorders.

Travelers who can't face the walk back to their rooms in Southwark or Leadenhall might like to bed down here for the night. Tower Hill and St Katharine's will be peaceful and well provisioned with wine and food, much of it intercepted on the way to the Tower and the rest given to them by Londoners. The party will go on all night long.

THURSDAY, JUNE 14TH: THE KING AT MILE END AND STORMING THE TOWER

Today you will need to choose one of three locations to focus on in the morning: HIGHBURY, MILE END, or the TOWER OF LONDON.

The village of HIGHBURY lies to the the north of the city and is home to a number of estates owned by Treasurer Hales. From early morning, a very large band of rebels, led by JACK STRAW, will have been systematically torching their great houses and stone barns. This group will be joined by a large body of townspeople from St Albans, alerted to the rebellion in London, and heading south to join it. Straw will later gather them together and ask them to collectively swear the rebel oath of allegiance "To King Richard and the Commons." Note that it is a three-mile hike to Highbury from the city: leave through Aldersgate and head north along Aldersgate Road and then Goswell Road.

The main political action of the day will be taking place just east of the city on the open fields of MILE END. After a night-long war council, KING RICHARD has decided to continue with his strategy of appeasement, and will be sending out messages to every quarter of the city to meet him at Mile End this morning. He will be riding out with a small retinue, including the Earls of Warwick and Oxford, his half-brothers, Thomas and John Holland, Sir Aubrey de Vere, who will be carrying the king's sword, his leading soldiers Sir Robert Knolles and Sir Thomas Percy, and Mayor Walworth. QUEEN JOAN will be bringing up the rear in a whirligig. Look out for the London captain THOMAS FARRINGDON, who accosts the king on a number of occasions, grabs the reins of his horse, and calls for the execution of Treasurer Hales—"That false traitor the prior." If you want to beat the crowds, head out before dawn through Aldgate and get yourself to Mile End, where most of the rebel army and much of the London mob will be gathering. Look out for two enormous books looted in Essex from the library of Admiral Edmund de la Mare; they will be carried around the field on the prongs of pitchforks, serving as strange rebel totems.

The choreography of the morning will be as follows. A small number of REBEL LEADERS WILL APPROACH KING RICHARD, initially addressing him on bended knee. They will make a long list of requests and complaints, generally concerning conditions of serfdom and work, and calls for an end to feudal fines and manorial controls. After a certain amount of this Richard will say "Yes" to all that, and call for the mob to line up in two long ranks so that he can confer a new charter of liberties on them. The king, at this point, for no obvious good reason, will loudly proclaim that, in addition to the freedom charters, the rebels are now free to go and catch traitors all across the realm of England, whom they should then bring to him for trial according to the due process of the law. Quite how much of that everyone hears is a moot point. You will find that within minutes most rebels have decided either to go home or, ignited by the royal Command, are baying to go and catch traitors in London. The due process of law will have to wait.

The news that King Richard has asked the rebels to seek out traitors will reach the TOWER OF LONDON by around 9:30am, and a large number of rebels will arrive there soon after from Mile End. The Tower, although appearing impregnable, will be captured by the rebels over the next hour; we have reports of the drawbridge being voluntarily lowered, the guards abandoning their positions on sighting a Royal Standard of England amongst the mob, and sympathisers within the Tower opening secret doors to the rebels.

Visitors may join the STORMING OF THE TOWER, but bearing in mind how easy it is to get lost in its maze of corridors and rooms, and how unsavoury it is going to be, it might be better to sit this one out. If you do go in, you are not, on any account, to reveal the hiding place of HENRY BOLINGBROKE, son of John of Gaunt, who in eighteen years is going to be crowned Henry IV of England. Over the next couple of hours rebel forces will reappear with four important prisoners: Archbishop of Canterbury SIMON OF SUDBURY, Royal Treasurer SIR ROBERT HALES, the king's Sergeant-at-Arms JOHN LEGGE and the Franciscan friar WILLIAM APPLETON, physician to John of Gaunt. All will be beheaded on Tower Hill. Sudbury's execution will be particularly inept, requiring eight blows from an axe to sever his head from his body.

Look out later on in the morning for QUEEN JOAN escaping on a barge to join the rest of the royals at their bolthole, CASTLE BAYNARD, in the west of the city.

AFTERNOON AND EVENING

The day is going to turn ever more gruesome. The anti-foreigner feeling among the London mob, in particular, will be given free rein. The ITALIAN BANKERS gathered around Lombard Street will be battening down the hatches, but by far the worst fate awaits the FLEMINGS, or Flemish, who are concentrated in the Vintry Ward, in the lanes south of Trinity Street running down to the Thames. At the junction of Queen Street and Upper Thames Street, you will find the mob gathering. They will eventually break down the church doors of the St Martin in the Vintry, where more than forty Flemings will be hiding, and then they will behead them all. Another victim of the mob in the Vintry this afternoon will be the fabulously wealthy London merchant RICHARD LYONS, who will be snatched from his house and dragged through the streets of the Cordwainer Ward to the north before being beheaded on Cheapside. Look out for what one contemporary describes as a man "Very fair and large, with his hair rounded by his ears and curled, a little beard forked, a gown, girt to him down to his feet, of branched damasks, wrought with the lines of flowers, a large purse on his right side, hanging in a belt from his left shoulder: a plain hood about his neck, covering his shoulders and hanging back behind him."

The heads of a number of those decapitated at Tower Hill in the morning will make their way across town to the ELEANOR CROSS in the hamlet of CHARRING, where rebel leaders will be orchestrating their display. Archbishop Sudbury, his red mitre nailed to his head, will be on show, joined by Sir Robert Hales, Roger Legget, William Appleton, and the juror Robert Somenour, all of them atop high spears or lances. In the late afternoon you can follow a raucous TOUR OF THE FIVE HEADS to Westminster and then back to the city, finishing at London Bridge, where they will be left on display.

FRIDAY, JUNE 15TH: WESTMINSTER ABBEY AND THE DEATH OF WAT TYLER

Early birds and royal watchers might like to base themselves in Westminster from dawn on Friday morning. RICHARD IMWORTH, the keeper of the king's Bench Prison in Southwark, is

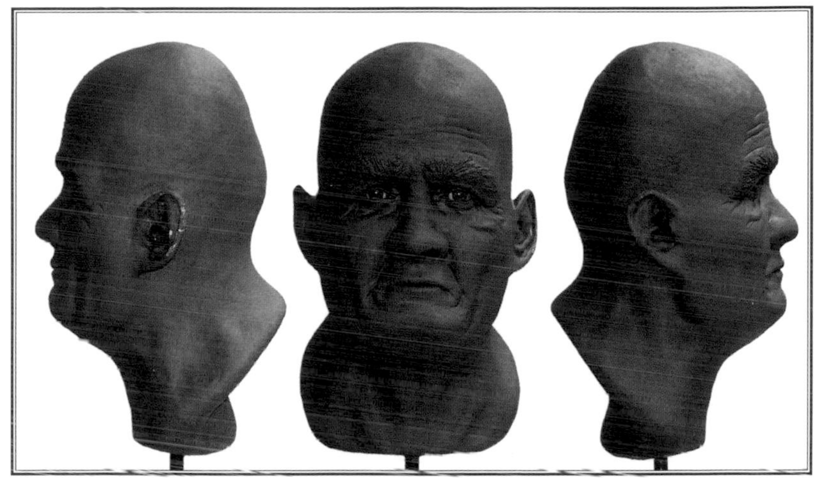

THIS REALLY IS SIMON OF SUDBURY—OR AT LEAST A RECONSTRUCTION OF HIS HEAD, CREATED BY A FORENSIC ARTIST FROM THE ARCHBISHOP'S SKULL, 600 YEARS AFTER IT PARTS COMPANY WITH HIS BODY.

one of the main targets of the rebels still abroad. He has fled his Southwark estates and taken refuge in WESTMINSTER ABBEY. At approximately 9am a large detachment of rebels will approach and enter the Abbey and will head for the High Altar, where Imworth will be hiding—and is taken captive. You can then follow the crowd back to Cheapside where Imworth will be beheaded in the early afternoon.

As you approach Bread Street on your return, look out for the unfortunate JOHN OF GREENFIELD, a valet who has apparently spoken well of John of Gaunt's physician William Appleton (who was beheaded yesterday). For this, he will be dragged by the mob down to Queenhithe and the banks of the Thames before being dragged back up to Cheapside, where he will be executed.

If you decide to stay in Westminster, await the ARRIVAL OF KING RICHARD. He and his retinue of over 200 will be met by a procession of the canons from St Stephen's Church, who will then accompany the king to Westminster Abbey for private prayers in the shrine of Edward the Confessor. Entry to the

Abbey will not be possible, and we recommend that you join the tail of the royal retinue when they depart an hour later for Smithfield.

EVENING

Visitors who choose to begin the day in the city centre should keep their ears open for the news that King Richard has called the rebels to a third meeting, which today is designated at SMITHFIELD—a large open space north of the city commonly used for fairs, festivals, and public executions. After lunch you will find a steady flow of people walking up to Newgate and Aldersgate—follow them. By around 5pm most of the remaining rebel forces in London will have gathered on Smithfield, and you will notice that WAT TYLER is once again on horseback and in command.

The royal retinue will arrive and assemble at some distance from the rebel army, in front of the hospital and the large stone Priory of St Bartholomew. You are unlikely to be able to get very close to the action to come, certainly not near enough to hear it, so a certain amount of interpreting body language will be required to make sense of events. You should be able to see Wat Tyler ride out towards Richard and his retinue. At a distance it is hard to see, but Tyler does appear to treat the king with a degree of contempt. There will be no bended knee this time; an over-firm handshake with the royal glove follows, and then you should see a long barracking during which Tyler purportedly outlines a UTOPIAN VISION OF ENGLAND in which much of the legal and spiritual apparatus of Plantagenet feudalism is dismantled. At some point here there will be an altercation between Mayor Walworth and Tyler, daggers will be drawn and used, and then, after further commotion, one of the royal retinue, RALPH STANDISH, will run Tyler through with a sword.

Confusion will reign all around, but keep your eyes on the king, who will call to the rebel army to follow him, and with his

MEDIEVAL COMIC BOOK CAPERS. RICHARD II WATCHES LONDON MAYOR RALPH STANDISH KILL WAT TYLER, AND TALKS TO THE PEASANTS AT THE SAME TIME.

bodyguard will ride off in the direction of Clerkenwell Field; after some indecision, much of the rebel force will follow him. Later on in the afternoon, MAYOR WALWORTH will return to Clerkenwell at the head of a small army of mercenary officers who will quickly surround the now demoralised and confused rebel forces. On Richard's order, and in the face of the newly arrived troops, you will see the rebel army slowly melt away, with much of the Essex contingent heading for home. If you can get close enough to the royal retinue at this point, you should be able to see Richard confer knighthoods on Mayor Walworth and three other aldermen: Nicholas Brembre, Robert Launde, and Ralph Standish.

DEPARTURE

We recommend that you make this your last port of call on this visit. Try to resist the urge to follow the Essex rebels home, or to explore the waves of rebellion and then counterterror that will ripple out across the country over the coming weeks from Bridgeport in Somerset to Cambridge and York.

First Battle of Bull Run

JULY 21, 1861 ✵ WASHINGTON D.C. & VIRGINIA, US

JOIN THE ELITES OF THE UNION AT THE first major battle of the American Civil War. Dine at WILLARD'S HOTEL, the centre of Washington D.C.'s social life, and smell the hubris; take a carriage to CENTREVILLE, Virginia, where the Union army of the Potomac is massing; sample the atmosphere on the CENTREVILLE KNOLL, where journalists, politicians, and the curious gather. For the intrepid, see and hear the battle from forward positions. But be warned: the devastating retreat and then rout of the Union army is to come... and you'll be just ahead of the pack skittering up the Warrenton Turnpike and back to Washington.

BRIEFING: EXPECTING A SWIFT VICTORY

In the early days of the American Civil War, the Union's politicians and press were calling for a quick and decisive victory over the rebel Confederacy in northern Virginia, paving the way for an assault on their capital in Richmond.

Under political pressure, and with some reluctance, given the untried nature of his forces, GENERAL IRVING MCDOWELL manoeuvred the UNION ARMY OF THE POTOMAC, 35,000 strong, southwest of Washington. There, the CONFEDERACY FORCES led by PIERRE GUSTAVE TOUTANT BEAUREGARD had massed at Manassas railway junction, on the far side of the BULL RUN RIVER.

The Washington political elite thought they were going to send the Confederacy packing at BULL RUN. Hundreds of civilians traveled the thirty miles from the capital to join General McDowell's army around the hamlet of CENTREVILLE, expecting a swift victory.

It won't quite work out that way. McDowell, due to a mixture

A UNION ARTILLERY UNIT ON THE BATTLEFIELD OF BULL RUN.

of poor scouting, inexperience, and bad luck, will find himself unable to concentrate his larger forces at the decisive moments of the battle. In particular, the great flanking movement of two of his divisions on the west of the battlefield will be slowed to a snail's pace by treacherous conditions. By contrast, Beauregard will be able to turn the tide of the battle when reinforcements arrive by rail. When the Union's line breaks at around 5pm, the discipline of this very inexperienced and now brutalised army will crack. Thousands of men will pour back to Centreville, and when harried by Confederate artillery and cavalry, it will become a rout.

Both armies are desperately naive and ill coordinated. All are shocked by the scale of the slaughter and by the harsh psychological and emotional conditions of battle. After Bull Run, it will become clear to all that the war will be long, gruelling, and costly.

THE TRIP

You will be arriving in WASHINGTON D.C. in late afternoon, Saturday, July 20, 1861. The weather will be warm and the air humid as you step out onto the southeast corner of the junction of PENNSYLVANIA AVENUE AND 14TH STREET. Please return here by midnight on Monday, July 22, for departure.

Washington, despite its apparent grandeur, is a new, small, and desperately underdeveloped city. Garbage disposal and sanitation are particularly poorly provided for—a situation made worse by the marshy and mosquito-ridden landscape on which the new Capitol has been constructed. Sniff the air. The rotten aromas of the city's putrid drainage canal, a quarter of a mile away, will be unmistakable.

Directly opposite you, however, is the beautiful curving corner of WILLARD'S HOTEL, a five-storey Beaux Arts concoction that serves as the capital's most prestigious visiting address. Home to President Lincoln before his inauguration earlier this year, and the site of the last desperate and failed peace conference before the war, it is now the centre of Washington's political and social life. Its salons and dining rooms are filled with politicians, lobbyists, senior military officers, journalists, and businessmen, not to mention, as the writer Nathaniel Hawthorne, who will be visiting Willard's in 1862, put it, "office seekers, wire pullers, inventors, artists, attachés of foreign powers, long-winded talkers, contractors and railway directors." Note the small patches of smoke and water damage on the building's walls. Just two months prior to your arrival, a fire in a small barber's shop next door threatened to engulf the hotel. A battalion of New York firemen, enlisted in the Union army as the 11th New York Zouaves (whom you will be meeting later), came to the rescue, forming human pyramids to carry water and hoses.

This is a good opportunity to take a look at the WHITE HOUSE just to your west and the WASHINGTON MONUMENT, five minutes' walk due south along 14th Street. CAPITOL HILL is a few minutes' ride by carriage or the regular omnibus service southwest along Pennsylvania Avenue.

You may be surprised by the architectural mayhem of the street. Broad, and lined with ailanthus trees, it is populated by a mixture of grand marble town houses, official buildings, rough wood-plank commercial premise, and tumbledown plots. The Capitol building itself is undergoing major work, its roofline covered with half-finished marble towers, ironwork, and cranes; the old dome on the building is being supplanted by a new and much bigger dome.

When you are ready to eat, head back to Willard's, walk into the lobby, and look for the dining rooms—reservations are not necessary. We recommend that you eat heartily while you can, as dining at Bull Run will be a much more haphazard affair. The menu features baked pike in claret sauce, roasted leg of mutton with capers, and broiled quails. For those with a sweet tooth, try the Lady Cake, cream pies, and jelly tart. As you will see from your fellow diners, PICNIC HAMPERS, SANDWICHES, AND BEER can also be ordered here for your journey.

We recommend you call for your carriage no later than midnight, and preferably much earlier. Those who wish to see the Union army 's pre-dawn mobilisation should be gone by 7pm.

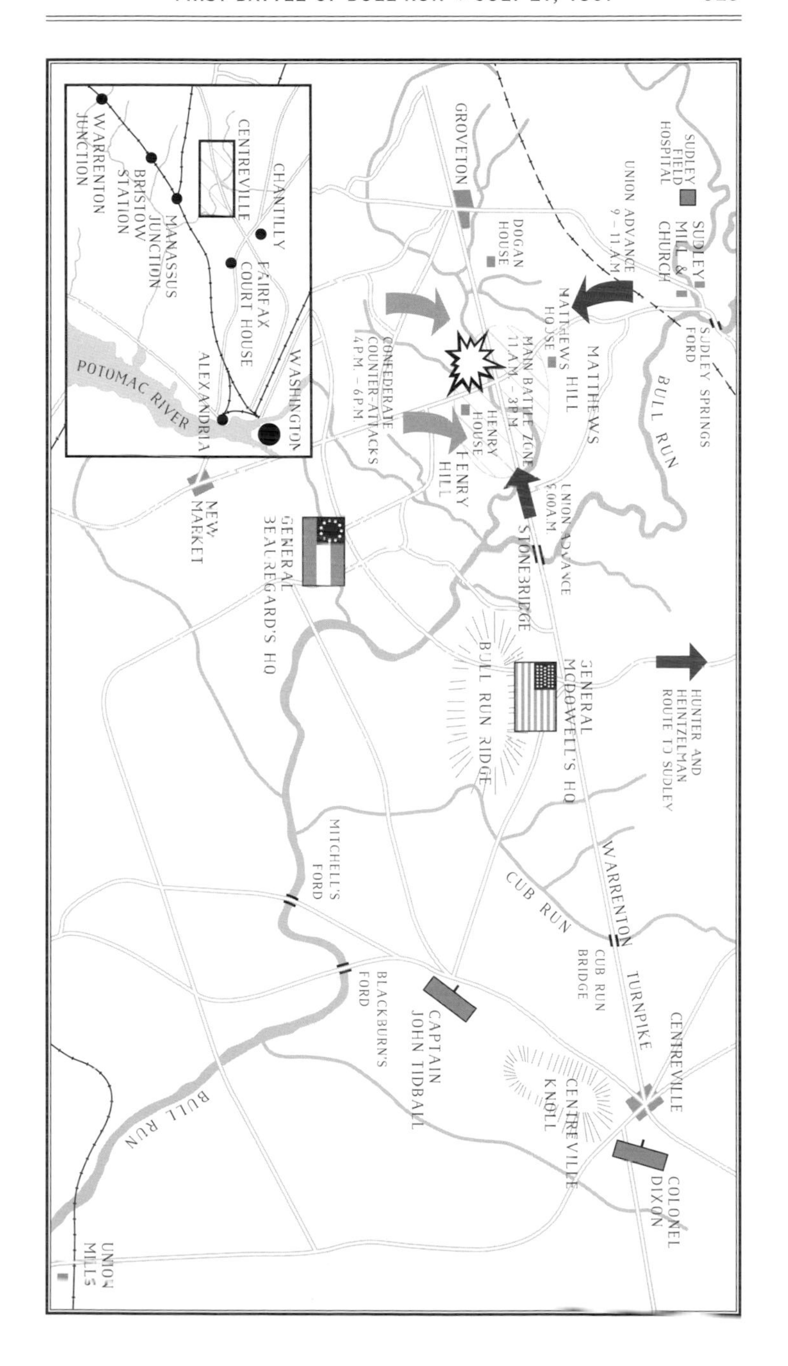
SUDLEY FIELD HOSPITAL
SUDLEY MILL & CHURCH
UNION ADVANCE 9 – 11 A.M.
GROVETON
DOGAN HOUSE
SUDLEY FORD
SUDLEY SPRINGS
BULL RUN
MATTHEWS HILL
MATTHEWS HOUSE
MAIN BATTLE ZONE 11 A.M. – 3 P.M.
HENRY HOUSE
HENRY HILL
CONFEDERATE COUNTER-ATTACKS 4 P.M. – 6 P.M.
UNION ADVANCE 6:00 A.M.
STONE BRIDGE
NEW MARKET
GENERAL BEAUREGARD'S HQ
HUNTER AND HEINTZELMAN ROUTE TO SUDLEY
GENERAL MCDOWELL'S HQ
BULL RUN RIDGE
WARRENTON TURNPIKE
CUB RUN
CUB RUN BRIDGE
MITCHELL'S FORD
BLACKBURN'S FORD
CAPTAIN JOHN TIDBALL
CENTREVILLE
CENTREVILLE KNOLL
COLONEL DIXON
BULL RUN
UNION MILLS
CHANTILLY
CENTREVILLE
FAIRFAX COURT HOUSE
WASHINGTON
ALEXANDRIA
POTOMAC RIVER
MANASSUS JUNCTION
BRISTOW STATION
WARRENTON JUNCTION

CENTREVILLE

Your carriage will be taking you west through GEORGETOWN and then south across the Potomac River before heading for CENTREVILLE in Prince William County, Virginia; it is about thirty miles—a seven-hour drive. Centreville is a tiny place, tired-looking at the best of times. On the morning of July 21st, you will find the town a great, ragged military encampment.

To its east are three UNION DIVISIONS commanded by COLONELS HUNTER, HEINTZELMAN, and MILES. To the west, about half a mile from town, is the division led by BRIGADIER GENERAL TYLER. The two encampments are connected by the Warrenton Turnpike, which serves as the town's main street. There are shutterboard houses, a small hotel, and a chapel, but little to trouble the visitor. The Union's commander-in-chief, GENERAL MCDOWELL, is camped close to the town on the west side. You might, while wandering near the camp, hear the general vomiting; he will be struck by a violent episode of food poisoning this evening.

Around 2:30am you will begin to hear the army stir: expect bugles, drum rolls and beats, and hundreds of campfires bursting into flame. Breakfast for most of the troops will be grim boiled coffee, pressed beef, hard-tack biscuits, and tobacco. Wagons will creek into action, horse harnesses will jangle, troops will be checking their weapons, and then there will be a rising crescendo of barked orders and movement. It will already be a pleasant temperature and will be considerably warmer by early morning. At the peak of the battle later in the day, it will be searing and many soldiers on both sides will suffer from severe dehydration. Now is the moment to make sure you have plenty of water supplies.

UNIFORMS

As the army mobilises, you will notice that there is no single, common uniform among the troops. Although the UNION has broadly opted for NAVY BLUE and the CONFEDERACY for GREY, there will be plenty of soldiers in both colors or the opponents' colors, for neither

THE ZOUAVE LOOK, AS MODELLED BY A UNION SOLDIER.

side has established centralised quartermasters' operations or imposed a common uniform. This sartorial irregularity is made worse by the fact that regiments are often kitted out by their city or state patrons, and officers have to buy their own uniforms. In any case, there will be so much dust, smoke, and mud at Bull Run later in the day that it will often be impossible to know who is who.

Two distinctive uniforms to look out for, however, are:

11TH NEW YORK INFANTRY (a.k.a. *Ellsworth* or *First Fire Zouaves*). The Zouaves are distinguished by red cuffs and trim on their navy jackets, as well as a red fireman's undershirt. They will be wearing a mixture of blue and red fezes and blue and red kepis.
14TH NEW YORK MILITIA (a.k.a. *14th Brooklyn* or *Red Legged Devils*). The Devils team their blue chasseur jackets with red pantaloons, a red-and-blue-banded kepi, and havelocks—white cloth-neck and head covers.

The ARMY'S MANOEUVRES, which will begin at about 4:30am, are simple in design but, as you will see, enormously complicated and fraught in practice. HUNTER AND HEINTZELMAN'S DIVISIONS—nearly 12,000 men, are camped equidistant from the turnpike but start marching simultaneously, leading to a terrible bottleneck of troops, horses, and equipment on the main street of Centreville. Worse, by the time the leading columns reach the west side of town, they will butt up against the rear of TYLER'S DIVISION still making its way down the Warrenton Turnpike to the STONE BRIDGE, having been delayed by fierce skirmishing along the path. Around 5:30am the jam will be cleared when General McDowell and his staff put in an appearance at the rickety wooden CUB RUN BRIDGE and order Tyler's remaining troops to get off the road and let Hunter and Heintzelman's divisions through.

For those visitors who want to experience SUDLEY FIELD HOSPITAL (see below), join the rear of Heintzelman's division as it turns off the Warrenton Turnpike a few hundred yards beyond the Cub Run Bridge. The march to Sudley Ford North and then back south through Sudley to the western edge of MATTHEWS HILL should take about five hours. Another 200 yards along the turnpike, you will find a blacksmith's, which has been turned into GENERAL MCDOWELL'S FORWARD FIELD HQ.

THE CENTREVILLE KNOLL

The GRASSY KNOLL to the south of Centreville is the main point at which civilian visitors to the battle will gather during the day. As you arrive in the early morning, a small number of journalists, politicians, and their retinues will already have staked out positions. Over the next five hours, numbers will swell as carriages, city hacks, and wagons roll in, joined by more visitors on horse or on foot. Amongst the early arrivals to look out for are WILLIAM HOWARD RUSSELL, correspondent for the *Times* of London, and the photographer MATHEW BRADY, identifiable by his straw hat, long linen duster, and the large wooden box (his camera) strapped to his back. From Capitol Hill, look out for SENATORS HENRY WILSON (Massachusetts); ZACHARIAH CHANDLER (Michigan); BEN WADE (Ohio); JAMES GRIMES (Iowa); JIM LANE (Kansas); LAFAYETTE FOSTER (Connecticut); and from the House of Representatives, ALFRED ELY (New York); SCHUYLER COLFAX (Indiana); and ELIHU WASHBURN and ISAAC ARNOLD (Illinois). Senator Wilson and his servants will be handing out freshly made sandwiches to troops for much of the morning. Do try to stay clear of his buggy and horses; they will be struck by a Confederate shell late in the afternoon, forcing the senator to escape back to Washington on a mule.

Other characters passing through the knoll will include JUDGE DANIEL MCCOOK of Ohio and the leading abolitionist W. P. THOMASSON, who is notable for his extremely high black-silk top hat. Later in the day he will pick up a rifle and join the 71st New York in battle, as will CONGRESSMAN OWEN LOVEJOY of Illinois. Two young Bostonians with a cart will pull up looking for the corpse of their brother—a Union soldier killed in a skirmish at Blackburn Ford, just south of the knoll, three days earlier. There are at least two dozen women on the knoll as well, including mothers of soldiers, wives of senators, and PIE AND SELTZER SELLERS. Look out for MISS AUGUSTA FOSTER, the adopted daughter and regimental mascot of the 2nd Maine.

In actual fact you will see very little of the battle from here. Though the sound of cannon fire and the smell of gun smoke will be prevalent all day, the main action on MATTHEWS HILL and HENRY HOUSE HILL is almost five miles away. At various points during the day, especially late morning and lunchtime, Union officers will ride up with good news from the front line—"We've whipped them at all points"; please do try not to give the game away. If you do decide to spend all day here, we suggest that you and your carriage are on the road out of Centreville by 6:30pm before the rout makes horse-drawn traffic impossible.

The more adventurous among the crowd, however, especially the senior Republican senators and journalists, will be making

COLONEL DIXON MILES—HARD LIQUOR NOT DEPICTED.

their way closer to the battlefield from 9am, heading for the WARRENTON TURNPIKE or the BULL RUN RIDGE (see below). More cautious travelers who would, nonetheless, like to get away from the knoll should explore BLACKBURN'S FORD and follow COLONEL DIXON MILES.

On a ridge about a mile south and slightly west of Centreville on the road to Manassas, you will find CAPTAIN JOHN TIDBALL and his artillery battery. From here you will have a good view of Bull Run Creek and the narrow Blackburn's Ford at the bottom of the ravine. There will still be corpses and abandoned equipment here, left over from a skirmish on the previous Thursday. By lunchtime there will be a considerable crowd, with buggies and carts parked in an overflow field behind the artillery battery, and a throng around Tidball, questioning him, fruitlessly, about the progress of the battle.

For those able to track him on horseback, the peregrinations of COLONEL DIXON MILES make for an interesting afternoon. Placed in command of the reserves, the colonel can be found in the morning on the stoop of the CENTREVILLE HOTEL, which is serving as his HQ and as a field hospital. He will be sporting the bizarre but effective sun protection of two straw hats mashed together on his head, and he will already be a little the worse for wear on a combination of liquor and opiate-based medicines. Colonel Dixon Miles will be making his first visits to the artillery positions of Colonels Davies and Richardson in the morning, but the afternoon return encounters will be the most colorful. Richardson in particular will refuse to accept Miles's orders and will publicly accuse him of being drunk. Hang around long enough at Richardson's position, and General McDowell himself will eventually show up and relieve Colonel Miles of his position.

BULL RUN RIDGE

Unquestionably the best vantage point from which to see at least some of the battle is the BULL RUN RIDGE just to the south of the Warrenton Turnpike. You can walk here across country but we suggest that you get on the turnpike and walk west. By late lunchtime the main reporters and senior senators will have gathered here with a view of at least parts of HENRY HOUSE HILL and MATTHEWS HILL, where the main fighting will be concentrated. For the very intrepid, the best view of all can be gained another half a mile down the Warrenton Turnpike at the STONE BRIDGE over Bull Run.

Over the afternoon, JUDGE DANIEL MCCOOK will be lunching with one of his sons, serving that day on the battlefield, and tragically leaving with his corpse in his buggy after he is shot by a Confederate officer. CONGRESSMAN WASHBURN, irate throughout the day, will be heading out on his reconnaissance mission in the afternoon. As ever, do make sure you are clear of the area by 5:30pm, when it will be overrun by the Confederate cavalry. CONGRESSMAN ELY will be taken prisoner by an infantry brigade.

BATTLEFIELD MEDICINE

The battle, although a mere skirmish compared to the massive encounters yet to come in the American Civil War, will appear unbelievably bloody and horrific to participants and observers. By the end of the day there will be 460 Union deaths and 1,124 wounded, while the Confederacy will lose 387 men and suffer 1,587 injuries. Many of the dead and injured will end up at one of the field hospitals established on the day.

For the medically minded among our travelers, UNION FIELD HOSPITALS can be found at: the four-room farmhouse by Stone Bridge; the Lewis House to the north of this; and on the Warrenton Turnpike beyond Bull Run Ridge. However, by far the largest and busiest is at SUDLEY SPRINGS, the village through which Hunter and Heintzelman's divisions marched in the morning.

Later in the day Sudley Springs cannot be approached across the battlefield, so you will have to follow Hunter and Heintzelman's men on their pre-dawn march. Here you will find that the church will be rapidly converted into a field hospital. Its pews will be removed and placed in the oak grove next to the church, and hay will be spread on the floor as improvised operating tables are set up inside.

The first AMBULANCE WAGONS, with their distinctive white-canvas covers, will begin arriving from 10:30am, blood dripping from the carts and gathering in a puddle outside the church door. Over the course of the afternoon, the medics will commandeer two nearby houses and a wheelwright's shop, but the churchyard will still fill to overflowing with the dead, the dying, and the wounded. Alongside dressing wounds, staunching cuts, and setting broken bones, the surgeons will be busy with amputations in an effort to save injured soldiers. Anaesthetics available include brandy, morphine, and chloroform, but we do warn travelers that the sounds of the hospitals are even more fearsome than the sights and smells.

THE BATTLE

From the vantage points of either the Stone Bridge or Bull Run Ridge, visitors will be able to glimpse only elements of the battle: the Union troops of TYLER'S DIVISION pouring across the Stone Bridge in mid-afternoon, skittish lines of troops crossing fields, plumes of smoke from the copses that dot the hillsides. Nonetheless, from these limited signals you should be able to put together the key choreography.

In the morning—from around 9:30am to 11:30am—Union and Confederate forces will clash on MATTHEWS HILL. You should be able to make out BRIGADIER GENERAL NATHAN EVANS'S SEVENTH BRIGADE shift position from in front of the Stone Bridge to the west side of Matthews Hill, where the main body of Union troops is beginning to arrive via Sudley Ford. Be alert to the advance of COLONEL SHERMAN'S 2ND BRIGADE, over Farm Ford a few hundred yards north of Stone Bridge. This pincer movement from the Union forces will, by around 11:30am, force the Confederate lines on Matthew's Hill to break. Visitors will be able to see them make their RETREAT south onto Henry House Hill.

Over the next hour or so, expect a steady ARTILLERY DUEL between the Union forces now massing on Matthews Hill and the reassembling Confederate lines on Henry House Hill.

IF THE BATTLE LOOKS LIKE THIS TO YOU (AND YOU'RE LOOKING HERE AT A UNION ATTACK), THEN YOU'RE PROBABLY TOO CLOSE. PLEASE RETURN TO THE BULL RUN RIDGE.

Between 1pm and 3pm, repeated UNION ATTACKS will be beaten off, including the now-famous, but probably apocryphal, moment when Virginian COLONEL JACKSON's regiment stands firm at the centre of the Confederate lines "like a stonewall."

It will not be immediately apparent from your vantage points, but after 3:30pm the TIDE OF BATTLE WILL BEGIN TO TURN. The Union's troops, many of them completely exhausted by their long march and lack of food and water, are beginning to wilt. At the same time a large number of CONFEDERATE REINFORCEMENTS will arrive from the Shenandoah Valley by train. When they are thrown into the battle, the UNION LINES WILL BREAK, first on the CHINN RIDGE on the eastern edge of the battlefield and then in the centre at the foot of Henry House Hill. By 5:00pm the Union army will be in HEADLONG RETREAT. Some will head north back to Sudley, but above all they will be flowing east over the Stone Bridge and back up the Warrenton Turnpike.

FEDERAL CAVALRY AT SUDLEY FORD AFTER THE BATTLE.

SENATORS ZACHARIAH CHANDLER AND BENJAMIN WADE.

THE RETREAT

The first signs of THE RETREAT will become apparent on Bull Run Ridge around 5pm. Fleeing soldiers will cry, "Turn back, turn back, we're whipped!" At this, expect SENATOR ZACHARIAH CHANDLER to respond by attempting to singlehandedly block the road, demanding that soldiers stop retreating. SENATOR BEN WADE will train a rifle on his troops, but in vain. It is now imperative that you get beyond the CUB RUN BRIDGE, a mile up the road, by 6pm; after that, Confederate fire will reach the bridge and induce the greatest moment of panic. A wagon on the bridge will swerve and turn over, blocking the path and forcing the streaming mass of people into the creek. At this point the retreat will become a ROUT. Soldiers will abandon their weapons, haversacks, and even uniforms, leaving a trail of ammunition boxes, hay bales, and bags of oats. Shells will be bursting overhead. Troops will clamber aboard ambulances, wagons, captured mules, and stray horses. Officers and men alike will be running pell-mell for Centreville, yelling with rage and fear; it is best not to get in their way.

Safety lies another mile up the turnpike, where COLONEL BLENKER'S GERMAN BRIGADE—three infantry regiments, plus cannon—will be covering the Union retreat and enforcing a modicum of order on the road. Some troops will be heading home, some will be heading to Washington, most will go to their camps in Centreville and grab what they can before beginning the thirty-mile walk back to the capital. General McDowell will order a COMPLETE RETREAT at midnight. If you can't get a carriage or a mule and have to walk, you might want to try to hook up with the Rhode Islanders, who will maintain the best morale and the best singing on the way back. It may be very gruesome; expect the walking wounded to include recent amputees, and soldiers with severed tongues and holes in their thighs and scrotums. As with the rest of the day, fresh water remains much in demand and in desperately short supply.

Past, Present & Future Reading

Alongside their own on-the-spot investigations and observations, WAC TIME INSPECTORS have made use of work by scholars and participants unable to access Chronoswoosh™ technologies. We can attest to the fidelity of the following books despite fixity in their own time–space continuum. They are all recommended pre-holiday reads—but should, of course, on no account be carried in your hand luggage.

The Field of the Cloth of Gold

Glenn Richardson ✻ *The Field of Cloth of Gold* (2013).

The St Louis World's Fair

S. Brownell (ed.) ✻ *The 1904 Anthropology Days and Olympic Games: Sport, Race, and American Imperialism* (2008).

G. R. Matthews ✻ *America's First Olympics: The St Louis Games of 1904* (2005).

N. J. Parezo & D. D. Fowler ✻ *Anthropology Goes to the Fair: The 1904 Louisiana Purchase Exposition* (2007).

VE Day

Russell Miller | *VE Day: The People's Story* (2007).

Woodstock Festival

James E. Perone ✻ *Woodstock: An Encyclopaedia of the Music and Art Fair* (2005).

Joel Makower ✻ *Woodstock: An Oral History* (1989).

The Boston Tea Party

Benjamin L. Carp ✻ *Defiance of the Patriots: The Boston Tea Party and the Making of America* (2010).

Harlow Giles Unger ✻ *American Tempest: How the Boston Tea Party Sparked a Revolution* (2011).

The Women's March on Versailles

George Rudé ✻ *The Crowd in the French Revolution* (1959).

Simon Schama ✻ *Citizens: A Chronicle of the French Revolution* (1989).

The Fall of the Berlin Wall

Mary Elise Sarotte ✻ *The Collapse: The Accidental Opening of the Berlin Wall* (2014).

The 235th Olympiad

Nigel Spivey ✻ *The Ancient Olympics: A History* (2005).

Shakespeare's Globe

Peter Ackroyd ✻ *Shakespeare: The Biography* (2005).

James Shapiro ✻ *1599: A Year in the Life of William Shakespeare* (2005).

Golden Age of Hollywood

Simon Louvish ✻ *Cecil B. DeMille: A Life in Art* (2008).

Scott Eyman ✻ *Empire of Dreams: The Epic Life of Cecil B. DeMille* (2010).

The Birth of Bebop

Stanley Crouch ✻ *Kansas City Lightning: The Rise and Times of Charlie Parker* (2013).

Ira Gitler ✻ *Swing to Bop: An Oral History of the Transition in Jazz in the 1940s* (1985).

Robin D.G. Kelley ✻ *Thelonious Monk: The Life and Times of an American Original* (2010).

The Beatles in Hamburg

Mark Lewisohn ✻ *The Beatles: All These Years—Vol. 1, Tune In* (2013).

The Rumble in the Jungle

Norman Mailer ✻ *The Fight* (1975).

In Xanadu with Marco Polo

Marco Polo ✻ *The Travels* (1299).

John Man ✻ *Xanadu: Marco Polo and Europe's Discovery of the East* (2009).

Captain Cook's First Epic Voyage

Peter Aughton ✻ *Endeavour: The Story of Captain Cook's First Great Epic Voyage* (2002).

James Cook ✻ *The Journals of Captain Cook* (1768–71).

Frank McLynn ✻ *Captain Cook: Master of the Seas* (2011).

The Eruption of Vesuvius

Mary Beard ✻ *Pompeii: The Life of a Roman Town* (2008).

The Peasants' Revolt

Judith Barker ✻ *1381: The Year of the Peasants' Revolt* (2014).

Dan Jones ✻ *Summer of Blood: The Peasants' Revolt of 1381* (2010).

First Battle of Bull Run

David Detzer ✻ *Donnybrook: The Battle of Bull Run, 1861* (2004).

PHOTOGRAPHY CREDITS

Every effort has been made to identify and contact copyright holders of images used in this book. If anyone has been omitted, they are asked to kindly contact the publisher so we can correct these details.

Francis I by Jean Clouet (p.8), Louvre Museum; *Henry VIII at the English Camp* (p.10–11), Royal Collection Trust/Her Majesty Queen Elizabeth II; *Henry VIII after Hans Holbein* (p.15), Walker Art Gallery. *V for Victory* (p.47), Picture Post/Getty Image; *Dancing in the streets* (p.52), Photo12/UIG/Getty Images; *Piccadilly Circus* (p.54) and *Churchill, Attlee, and Bevin* (p.57), Keystone/Getty Images. *Michael Lang and Artie Kornfeld* (p.62), Bill Eppridge/Life/Getty Images; *Breakfast* (p.67), John Dominis/Life/Getty Images; *Undress code* (p.71), Silver Screen/Movipix/Getty Images; *Swami Satchidananda* (p.74), Mark Goff/WikiCommons; *Country Joe* (p.77), Jason Laure/Woodfin/Getty Images; *Lost girl* (p.80), Three Lions/Hulton Archive/Getty Images. *Tea tax cartoon* (p.93) and *Mohawks* (p.99), Hulton Archive/Getty Images. *Tardivet and Miomandre* (p.122), Bibliothèque Nationale de France. *Berlin Wall "Death Strip"* (p.125), Thierry Noir/WikiCommons; *Brandenburg Gate* (p.131), Suc Ream/WikiCommons; *Rostropovich* (p.137), L. Emmett Lewis Jr. © Stars and Stripes. *Pankration scene* (p.156), Marie-Lan Nguyen/Wikimedia Commons/CC-BY 2.5. *Charlie Parker* (p.203), Gilles Petard/Getty Images; *Lindy Hoppers at the Savoy* (p.206), Charles Peterson/Hulton Archive/Getty Images; *Outside Minton's Playhouse* (p.211), William Gottlieb/Redferns/Getty Images. *Fab Five* (p.217), Astrid Kirchherr/K&K/Redferns/Getty Images; *The Beatles at the Top Ten Club* (p.224), Ellen Piel/K&K/Redferns/Getty Images; *The Beatles at the Star Club* (p.229), Ulf Kruger/K&K/Redferns/Getty Images. *Mobutu introduces Foreman and Ali* (p.232), George Walker/Liaison/Getty Images; *Foreman in training* (p.238), Neil Leifer/Sports Illustrated/Getty Images; *Ali prepares* (p.241), Stringer/AFP/Getty Images; *Ali throws the Big One* (p.245), The Ring Magazine/Getty Images. *Kublai Khan* (p.257), Dea/Getty Images. *Tahitian women* (p.277), Time Life Pictures/Mansell/Life/Getty Images; *The Endeavour* (p.284), SSPL/Getty Images. *Up Pompeii!* (p.296), Art Media/Print Collector/Getty Images. *John Ball* (p.303), British Library Board. *Union artillery unit* (p.322), *Zouave uniform* (p.327), *Colonel Dixon Miles* (p.330), *Union attack* (p.333), *Federal Cavalry at Sudley Springs* (p.334), *Senators Zachariah Chandler and Benjamin Wade* (p. 335), all Library of Congress.

All MAPS by Magnetic North.